T0297449

Studies in Computational Intelligence

Volume 633

Series editor

Janusz Kacprzyk, Polish Academy of Sciences, Warsaw, Poland
e-mail: kacprzyk@ibspan.waw.pl

About this Series

The series "Studies in Computational Intelligence" (SCI) publishes new developments and advances in the various areas of computational intelligence—quickly and with a high quality. The intent is to cover the theory, applications, and design methods of computational intelligence, as embedded in the fields of engineering, computer science, physics and life sciences, as well as the methodologies behind them. The series contains monographs, lecture notes and edited volumes in computational intelligence spanning the areas of neural networks, connectionist systems, genetic algorithms, evolutionary computation, artificial intelligence, cellular automata, self-organizing systems, soft computing, fuzzy systems, and hybrid intelligent systems. Of particular value to both the contributors and the readership are the short publication timeframe and the worldwide distribution, which enable both wide and rapid dissemination of research output.

More information about this series at http://www.springer.com/series/7092

Kiran Kumar Ravulakollu
Mohammad Ayoub Khan
Ajith Abraham

Editors

Trends in Ambient Intelligent Systems

The Role of Computational Intelligence

 Springer

Editors
Kiran Kumar Ravulakollu
Department of Computer Science
  and Engineering, School of Engineering
  and Technology
Sharda University
Greater Noida
India

Ajith Abraham
Machine Intelligence Research Labs
  (MIR Labs)
Auburn, WA
USA

Mohammad Ayoub Khan
Ministry of Communications
  and Information Technology
Centre for Development of Advanced
  Computing
Noida
India

ISSN 1860-949X   ISSN 1860-9503   (electronic)
Studies in Computational Intelligence
ISBN 978-3-319-30182-2   ISBN 978-3-319-30184-6   (eBook)
DOI 10.1007/978-3-319-30184-6

Library of Congress Control Number: 2016932000

Printed on acid-free paper

This Springer imprint is published by SpringerNature
The registered company is Springer International Publishing AG Switzerland

Preface

Intelligence has always been a fascinating dimension that is 'simple to understand but complex to achieve'. For understanding intelligence various pathways are designed and defined along the progress of research. However, engineering intelligence into machines for communication, cooperation and coordination is always challenging. In this context, artificial intelligence can be defined as "a way through which machine can be made to understand environment, logically reason the information and learn for acting better". Though the principle of intelligence is clear and convincing, achieving total intelligence is yet a dream come true. Over the years, many domains are developed for specialization and nourishment. Though this area has spread its roots through various specializations in addressing intelligence, it always suffers from limitations on selective applicability.

It is observed that technological advancements are much more effective when they are addressing day-to-day environments. The dynamic nature of these environments sets the challenges to technologies for their feasibility, stability and adaptability. The motivation for this book is to demonstrate the success of ambient intelligence in providing solutions for daily needs of humanity in ways possible. This book addresses implications of ambient intelligence in areas of domestic living, elderly care, robotics, communication, philosophy and so on.

Ambient intelligence, an amalgamation of need, convenience, behaviour, technology and intelligence is applied with greater significance over all sorts of living. Ambient intelligence is also an attempt to incorporate so-called "Hi-Tech" infrastructure into day-to-day life with an aim to improve quality and standard of living. The objective of this edited volume is to justify and enrich the philosophy that ambient intelligence is a boon to humanity with conceptual, philosophical, methodical and applicative understanding. The book also aims to schematically demonstrate developments in the direction of augmented sensors, embedded systems and behavioural intelligence towards Ambient Intelligent Networks or Smart Living Technology. The book contains chapters in the field of Ambient Intelligent Networks, which received highly recommended feedback during the review process. This book contains research work, with in-depth state of the art from augmented sensors, embedded technology and artificial intelligence along with cutting-edge research and development of technologies and applications of Ambient Intelligent Networks. For readers of relevant research communities and individuals, this book is intended to introduce ideas, methods, technologies of the future development of humanity, Science and Technology.

Kiran Kumar Ravulakollu

Contents

About the Editors

Dr. Kiran Kumar Ravulakollu is working with Sharda University of India as Assistant Professor with interests in sensory networks, robotics, biologically inspired multimodal behaviour modelling, ambient intelligence, and sensory network design along with visual and auditory information processing. Other areas of interests include integrated areas of agent navigation, human–computer interaction and neural networks. He has more than five years of research experience in Hybrid Intelligent Systems area at Centre for Hybrid Intelligent Systems, University of Sunderland, UK. He received his Ph.D. degree from University of Sunderland, UK, and Bachelor of Technology Degree from Jawaharlal Nehru Technological University, India. He contributes to research communities by various volunteer activities. He has been publishing with internationally reputed publishers like Springer, Elsevier through journals, conference and book chapters. He has served on the editorial/reviewer board for journals like IJACR, Hindawi Publishing Corporation and JECE along with conferences like ICECCS-2013, ICCTD-2011. He was organizing committee member for ncaf-2009 conference held in UK. He has served as conference chair for international conference of SocProS-2013. He is a member the professional bodies IEEE, INNS and IACSIT. He can be reached at kiran007.r@gmail.com.

Dr. Mohammad Ayoub Khan is Associate Professor, Department of Computer Science and Engineering, School of Engineering and Technology, at Sharda University, Greater Noida, India, with interests in radio frequency identification, microcircuit design, and signal processing, NFC, front end VLSI (Electronic Design Automation, Circuit optimization, Timing Analysis), placement and routing in network-on-chip, real time and embedded systems. He has more than 11 years of experience in his research area. He has published more than 60 research papers in reputed journals and international IEEE conferences. He contributes to the research community by various volunteer activities. He has served as conference chair in various reputed IEEE/Springer international conferences. He is a senior member of the professional bodies IEEE, ACM, ISTE and EURASIP society. He may be reached at ayoub@ieee.org.

Dr. Ajith Abraham received the Ph.D. degree in Computer Science from Monash University, Melbourne, Australia. He is currently the Director of Machine Intelligence Research Labs (MIR Labs), Scientific Network for Innovation and Research Excellence, USA, which has members from more than 100 countries. He has worldwide academic and industrial experience of over 23 years. He works in a multidisciplinary environment involving machine intelligence, network security, various aspects of networks, e-commerce, Web intelligence, Web services, computational grids, data mining and their applications to various real-world problems. He has numerous publications/citations (h-index 60) and has also given more than 70 plenary lectures and conference tutorials in these areas. Since 2008, he is the Chair of IEEE Systems Man and Cybernetics Society Technical Committee on Soft Computing and a Distinguished Lecturer of IEEE Computer Society representing Europe (since 2011). Dr. Abraham is a Senior Member of IEEE, the Institution of Engineering and Technology (UK) and the Institution of Engineers Australia (Australia), etc. He is the founder of several IEEE sponsored annual conferences, which are now annual events. More information at: http://www.softcomputing.net.

A Neurocognitive Robot Assistant
for Robust Event Detection

German I. Parisi and Stefan Wermter

Abstract Falls represent a major problem in the public health care domain, especially among the elderly population. Therefore, there is a motivation to provide technological solutions for assisted living in home environments. We introduce a neurocognitive robot assistant that monitors a person in a household environment. In contrast to the use of a static-view sensor, a mobile humanoid robot will keep the moving person in view and track his/her position and body motion characteristics. A learning neural system is responsible for processing the visual information from a depth sensor and denoising the live video stream to reliably detect fall events in real time. Whenever a fall event occurs, the humanoid will approach the person and ask whether assistance is required. The robot will then take an image of the fallen person that can be sent to the person's caregiver for further human evaluation and agile intervention. In this paper, we present a number of experiments with a mobile robot in a home-like environment along with an evaluation of our fall detection framework. The experimental results show the promising contribution of our system to assistive robotics for fall detection of the elderly at home.

1 Introduction

Falls represent a major concern in the public health care domain, especially among the elderly population. According to the World Health Organization, fall-related injuries are common among older persons and represent the leading cause of pain, disability, loss of independence and premature death [1]. Although fall events do not necessarily cause a fatal injury, fallen people may be unable to get up without

G.I. Parisi (✉) · S. Wermter
Department of Informatics, Knowledge Technology, University of Hamburg,
Vogt-Koelln-Strasse 30, 22527 Hamburg, Germany
e-mail: Parisi@informatik.uni-hamburg.de
URL: http://www.informatik.uni-hamburg.de/WTM/

S. Wermter
e-mail: wermter@informatik.uni-hamburg.de

© Springer International Publishing Switzerland 2016
K.K. Ravulakollu et al. (eds.), *Trends in Ambient Intelligent Systems*,
Studies in Computational Intelligence 633, DOI 10.1007/978-3-319-30184-6_1

assistance, thereby resulting in "long lie" complications such as hypothermia, dehydration, bronchopneumonia, and pressure sores [2]. Moreover, fear of falling has been associated with a decreased quality of life, avoidance of activities, and mood disorders such as depression (among fallers and non-fallers) [3].

As a response to increasing life expectancy, plenty of research has been done to provide technological solutions for supporting living at home and smart environments for assisted living. The motivation of assistive fall systems is the ability to promptly report a fall event and by this enhancing the person's safety perception and avoiding the loss of confidence due to functional disabilities. Recent systems for elderly care aim mostly to detect hazardous events such as falls and allow the monitoring of physiological measurements (e.g. heart rate, breath rate) with the use of wearable sensors to detect and report emergency situations in real time [4, 5]. Vision-based fall detection is currently the predominant approach due to the constant development of computer vision techniques that yield increasingly promising results in both experimental and real-world scenarios. Additionally, in the last half decade the advent of low-cost depth-sensing devices such as the Microsoft Kinect [6] and ASUS Xtion Live [7] has led to a great number of vision-based applications using depth information instead of, or in combination with, color information. In this setting, the use of machine learning and neural network approaches has been shown to be an appropriate methodology to achieve knowledge generalization of a set of training activities for the classification of unseen situations [8], and the detection of abnormal behaviors such as fall events in domestic environments [9].

Contrary to fixed sensors, mobile assistive robots may be designed to process the sensed information and undertake actions that benefit people with disabilities and seniors in a residential context. There exists an increasing number of ongoing research projects using assistive robotics in smart environments to provide tools for self-care, independence at home, and telematic diagnosis. Moreover, advanced robotic technologies may encompass socially-aware assistive solutions for interactive robot companions, able to support basic daily tasks of independent living and enhance user experience through human-robot interaction (e.g. dialogues and vocal commands). Recent studies support the idea that the use of socially assistive robots leads to positive effects on the senior's well-being in domestic environments [10]. On the other hand, the use of robotic technologies brings a vast set of challenges and technical concerns.

In this work, we introduce a humanoid robot assistant that monitors a person in a household environment and reports abnormal user behavior such as a fall event. The underlying motivation is that the robot keeps the person in the scene while he/she performs daily activities, thereby anonymously tracking the user's position, body posture, and motion characteristics. The processed visual information is fed into the neural system which is responsible for triggering alarms in case a fall is detected. Whenever a fall event occurs, the humanoid will approach the user and ask whether assistance is required. The robot will then take a picture of the scene that can be sent to the user's caregiver or relatives for telematic evaluation.

This chapter is organized as follows. In Sect. 2, we provide an overview on the state of the art in fall detection, in particular vision-based approaches using depth sensors and assistive robotics. In Sect. 3, we introduce our learning-based neural framework for detecting abnormal events. We show experiments in a home-like environment and an evaluation of the system for a person falling down or crawling. In Sect. 4, we present an assistive humanoid robot for detecting fall events in a domestic scenario. We first depict an overview of our system and then go into detail about the software, hardware, and the communication interface. We conclude in Sect. 5 with a discussion on open issues in fall detection, trends and challenges for assistive robots, and future work directions for *aging at home* systems.

2   Trends in Fall Detection

Broadly speaking, a fall detection system can be defined as an assistive service with the main goal to promptly report a fall event. From a technical perspective, this service represents a pervasively challenging task in real-world scenarios in terms of reliability and robustness, since it raises a vast set of issues and technological concerns. As reported by an extensive number of works in the literature, fall detection systems may be designed, implemented, and evaluated on the basis of a manifold of approaches using different types of sensing devices and methodologies to process the sensed information.

The purpose of this section is to provide a concise overview of the state of the art in fall detection technologies with a particular focus on vision-based approaches, the developing use of low-cost depth sensing devices for 3D tracking, and emerging technologies in assistive robotics for aging at home and telematic caregiving.

2.1   Fall Detection Systems

There seems to exist an agreed taxonomy in the literature that classifies fall detection systems into two main categories according to the type of sensor used to monitor the user: wearable-based and ambient-based approaches [11, 12].

Wearable-based approaches relate to the use of small electronic devices that can be worn by the user, for instance, on top of clothing or as accessories. The most extensively used wearable devices consist of accelerometers and gyroscopes attached to the body that measure the user's location and motion. There is a vast number of applications that use these measurements to evaluate the user's gait and balance, and assess the risk of a fall event [13–16]. In the last years, this trend has seen a significant boost due to the availability of low-cost sensors embedded in smartphones [17–20]. On the other hand, ambient-based approaches relate to the use of sensing devices deployed in the environment, thereby not requiring the user to wear any sensor. Fall detection systems of this kind, also referred to as non-intrusive and context-aware, may encompass a wide spectrum of sensor types such as cameras, microphones, pressure and floor sensors [21, 22].

We focus on the use of cameras for vision-based fall detection with increasingly promising results in both experimental and real-world scenarios. Lee and Mihailidis [23] presented a vision-based method with a ceiling camera for monitoring falls at home. The authors considered falls as lying down in a stretched or tucked position. The system accuracy was evaluated with a pilot study using 21 subjects consisting of 126 simulated falls. Personalized thresholds for fall detection were based on the height of the subjects. The system detected fall events with 77 % accuracy and had a false alarm rate of 5 %. Miaou et al. [24] presented a customized fall detection system using an omni-camera for capturing 360° scene images. Falls were detected based on the change of the ratio of people's height and width. Two scenarios were used for the detection: with and without considering user health history, for which the system showed 81 and 70 % accuracy respectively. Rougier et al. [25] presented a method for fall detection by analysing human shape deformation in video sequences. Falls were detected from normal activities using a Gaussian mixture model with 98 % accuracy. The overall system performance increased when taking into account the lack of significant body motion after the detected fall event. Liu et al. [26] detected falls considering privacy issues, thereby processing only human silhouettes without featural properties such as the face. A k-nearest neighbor (kNN) algorithm was used to classify the postures using the ratio and difference of a body silhouette bounding box. Recognized postures were divided into three categories: standing, temporary transitional, and lying down. Experiments with 15 subjects showed a detection accuracy of 84.44 % on fall and lying down events.

In a multi-camera scenario, Cucchiara et al. [27] presented a vision system with multiple cameras for tracking people in different rooms and detecting falls based on a hidden Markov model (HMM). People tracking was based on geometrical and color constraints and then sent to the HMM-based posture classifier. Four main postures were considered: walking, sitting, crawling, and lying down. When a fall was detected, the system triggered an alarm via SMS to a clinician's PDA with a link to live low-bandwidth video streaming. Experiments showed that occlusions had a strong negative impact on the system's performance. Hazelhoff et al. [28] detected falls using two fixed perpendicular cameras. The foreground region was extracted from both cameras and the principal components (PCA) for each object were computed to determine the direction of the main axis of the body and the ratio of the variances. Using these features, a Gaussian multi-frame classifier was used to recognize falls. In order to increase robustness and mitigate false positives, the position of the head was taken into account. The system was evaluated also for partially occluded people. Experiments showed real-time performance with an 85 % overall detection rate.

In contrast to the use of color cameras, Diraco et al. [29] addressed the detection of falls and the recognition of several postures with 3D information. The system used a fixed time-of-flight camera that provided robust measurements under different illumination settings. Moving regions with respect to the floor plane were detected applying a Bayesian segmentation to the 3D point cloud. Posture recognition was carried out using the 3D body centroid distance from the floor plane and the estimated body orientation. The system yielded promising results on synthetic data with threshold-based clustering for different centroid's height thresholds.

An enduring bottleneck for vision-based approaches is the segmentation of human shape from acquired 2D image sequences, which is often constrained in terms of computational effort and robustness to illumination changes. Recent research work has indicated a trend towards fall detection systems using 3D sensing devices for more accurate and efficient estimations of human motion and body posture.

2.2 3D Human Tracking

In the last half decade, the emergence of low-cost depth sensing devices such as the Microsoft Kinect [6] and ASUS Xtion Live [7] has led to a great number of vision-based applications using depth information instead of, or in combination with, color information. This prominent sensor technology provides depth measurements used to obtain reliable estimations of 3D human motion in cluttered environments, including a set of body joints in real-world coordinates and their orientations. As shown by a broad number of recent applications for human action recognition, this sensor trend represents a significant contribution to overcome a set of limitations related to traditional 2D sensors (e.g. RGB cameras), thereby increasing robustness under varying illumination conditions and reducing computational effort for motion segmentation and body pose estimation. Depth sensors have the additional advantage of avoiding privacy issues regarding the identity of the monitored person, since color information is not required at any stage. An extensive review of the depth sensor Kinect and its application to diverse research fields, e.g. action recognition and navigation, was presented by Han et al. [30].

A combination of computational efficiency, robustness to light changes in indoor environments, and lower cost factors have made fall detection systems using depth information increasingly popular in the research community. Rougier et al. [31] used 3D information from a depth sensor to estimate a person's centroid height and velocity relative to the ground plane. Thresholds on ground distance and velocity computed from training data were used to detect fall events also with occluded persons (e.g. fallen down behind a sofa). The system was evaluated on simulated falls and normal activities (e.g. walking, sitting down, crouching) with an overall success rate of 98.7 %. Planinc and Kampel [32] used depth information to compute a body axis that described the overall orientation of a person. Thresholds for similarity to the ground and the height were used to distinguish falls from other daily activities. The system was evaluated on a dataset of 72 video sequences containing 40 falls with accuracy of 95 % after eliminating tracking errors. In this approach, occlusions were not considered. Mastorakis and Makris [33] presented a depth-based fall detection system taking into account body velocity and inactivity periods. The velocity was measured on the basis of the contraction or expansion of a 3D bounding box built around the person's body. The detection algorithm was designed as a Boolean decision tree for distinguishing falls from other actions. Good results were obtained from different sensor perspectives (frontal, side) on a customized dataset.

Approaches using depth information in combination with machine learning and neural networks have shown to provide promising results. Zhang et al. [34] presented a depth-based system to recognize different types of falls, i.e. fall from a standing position and fall from a chair. Body features such as structure similarity and height were extracted from a kinematic model and fed to a hierarchical Support Vector Machine (SVM) classifier. Promising results for detecting falls from other three daily actions (i.e. standing, sitting on a chair or on the floor) were obtained on a dataset of 200 video sequences with different light conditions. Parisi and Wermter [9] presented a neural network approach to detect abnormal behaviors such as falling, fainting, and crawling while monitoring domestic daily actions. A self-organizing neural architecture was trained on a set of domestic actions (e.g. walking, sitting, picking up objects) from body features such as velocity and orientation. The system detected abnormal behavioral patterns not shown during the training phase in two different tracking scenarios with fixed and mobile depth sensors. Best results were obtained by automatically detecting and removing tracking errors.

In contrast to most of the approaches using the depth sensor positioned parallel to the horizontal surface, Gasparrini et al. [35] detected falls using a ceiling sensor. A segmentation algorithm was used to extract blobs in the scene and track human silhouettes on the basis of several anthropometric relations. Falls were detected for a tracked person under a threshold-based distance to the floor. Experiments showed promising results also for scenarios with more persons present in the top-view scene.

While the number of advantages introduced by low-cost depth sensors is significant in terms of body motion and posture estimation, these approaches lead to issues that may prevent them from operating in real-world environments. For instance, their operation range (distance covered by the sensor) is quite limited (between 0.8 and 5 m), as well as their field of view (see Table 1 for details), thereby requiring a mobile or multi-sensor scenario to monitor an extensive area of interest.

Table 1 ASUS Xtion Live sensor specifications [7]

Depth image size	VGA (640 × 480): 30 fps, QVGA (320 × 240): 60 fps
Field of view	58 H, 45 V, 70 D (horizontal, vertical, diagonal)
Distance of use	0.8–3.5 m
Dimensions	18 × 3.5 × 5 cm
Power consumption	Below 2.5 W
Interface	USB 2.0/3.0
Weight	227 g

2.3   Assistive Robotics

Mobile robots have been characterized by a constant development for "aging at home" scenarios. In contrast to fixed sensors, mobile assistive robots may be designed to process the sensed information and undertake actions that benefit people with disabilities and seniors in a residential context. In fact, the mobility of robots represents a big benefit for non-invasive monitoring of users, thereby better addressing fixed sensors' limited field of view, blind spots, and occlusions.

There has been an increasing number of ongoing research projects using assistive robotics in smart environments to provide tools for self-care, independence at home, and telematic diagnosis. Advanced robotic technologies may encompass socially-aware assistive solutions for interactive robot companions able to support basic daily tasks of independent living and enhance user experience through flexible human-robot interaction (e.g. dialogues, vocal commands). A number of experimental studies support the idea that the use of socially assistive robots implies positive effects on the seniors' well-being in domestic environments [10]. Examples of recent and current interdisciplinary research projects using interactive mobile robots for aging in place include: Cogniron (Cognitive Robot Companion) [36], LIREC: Living with robots and interactive companions [37], Hermes: Cognitive Care and Guidance for Active Ageing [38], KSERA (Knowledgable SErvice Robots for Aging) [39], GiraffPlus [40], ROBOT-ERA [41], and Accompany (ACceptable robotics COMPanions for AgeiNg Years) [42]. Despite different functional perspectives concerning elderly care and user needs (e.g. rehabilitation [39], robot companions [42]), there is a strong affinity regarding the intrinsic challenges and issues needed to operate these systems in real-world scenarios. In fact, the use of mobile robots may be generally combined with ambient sensors embedded in the environment (e.g. cameras, microphones) to enhance the agent's perception and increase robustness under real-world conditions. On the other hand, complementary research efforts have been conducted on the deployment of stand-alone mobile robot platforms able to sense and navigate the environment by relying exclusively on onboard sensors.

Specifically for fall detection, promising experimental results have been obtained by combining mobile robots and 3D information from depth sensors. This approach overcomes limitations in the operation range of sensors while preserving reduced computational power for real-time characteristics. Mundher and Zhong [43] proposed a mobile robot with a Kinect sensor for fall detection based on floor-plane estimation. The robot tracks and follows the user in an indoor environment, and can trigger an alarm in case of a detected fall event. The system recognizes two gestures to start and stop a distance-based *user-following* procedure, and three voice commands to enable/disable fall detection, and call for help. The robot is provided with a mobile phone to send notifications via SMS or emergency call if the user does not recover from a fall within five seconds. Volkhardt et al. [44] presented a mobile robot to detect fallen persons, i.e. a user already lying on the floor. The system segments objects from the ground plane and layers them to address partial occlusions. A classifier trained on positive and negative examples is used to detect object layers as a fallen human. Experiments reveal that the overall accuracy of the system is strongly dependent on the type of extracted features and the classifier.

Additional challenges conveyed by the use of mobile robots for detecting fall events regard the tolerance of noise in a moving sensor scenario [9], the robust tracking of occluded persons [45], and effective navigation strategies for following and finding people in domestic environments [46].

3   Learning-based Abnormal Event Detection

Despite extensive research efforts promoted by advanced computer vision techniques and recent low-cost sensor trends, the question remains open on how to better process extracted body features for effectively extrapolating the complex dynamics of actions and fall events exhibiting noise tolerance and robustness in real-world scenarios. Indeed, the vast majority of the presented algorithms rely on domain-specific thresholds to distinguish falls from other activities, often being unable to operate under real-world conditions. On the other hand, learning-based paradigms such as machine learning and neural networks represent prominent tools to achieve knowledge generalization in a set of training activities for the subsequent classification of unseen situations [8, 9, 47]. In this setting, a possible approach for fall detection consists in learning a set of normal actions from training data and subsequently detecting events that do not conform to the expected behavior.

In this section, we present our work on abnormal event detection based on unsupervised neural network learning. The system consists of a hierarchical neural architecture that learns a set of normal actions, e.g. walking, sitting, and picking up objects, captured by a depth sensor. After the training phase, the system will report novel behavioral patterns, e.g., fall event, as abnormal actions and trigger an alarm. To contrast tracking errors and sensor noise, the neural architecture is also responsible for automatically removing noisy samples from the extracted body features. We report a number of experiments in a home-like environment that show our system can detect fall events with high accuracy in real time.

3.1   Feature Extraction

The first stage of our system consists of the extraction of body action features from 3D motion information captured by a depth sensor. We estimate the position of a moving target based on a model of the human skeleton. In previous work [9], we used this skeleton-based representation to compute body centroids that describe actor-independent posture and motion features. Two centroids were estimated as the centers of mass that follow the distribution of the main body masses on each

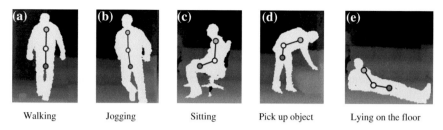

Fig. 1 Full-body representation for pose-motion extraction [47]. We estimate three centroids C_1 (*green*), C_2 (*yellow*) and C_3 (*blue*) for upper, middle and lower body respectively. We compute the segment slopes (θ^u and θ^l) to describe the posture with the overall orientation of the upper and lower body

posture. This technique extrapolates significant motion characteristics while maintaining a low-dimensional feature space and increasing tracking robustness for situations of partial occlusions. We then extended our model to describe more accurately articulated actions by considering three body centroids [47]: C_1 for upper body with respect to the shoulders and the torso; C_2 for middle body with respect to the torso and the hips; and C_3 for lower body with respect to the hips and the knees. Each centroid is represented as a point sequence of real-world coordinates $C = (x, y, z)$. We compute upper and lower orientations θ^u and θ^l given by the slope angles of the segments $\overline{C_1 C_2}$ and $\overline{C_2 C_3}$ respectively. As shown in Fig. 1, θ^u and θ^l describe the overall body posture as the overall orientation of the torso and the legs, allowing to capture significant posture configurations of actions such as walking, sitting, picking up and lying down on the floor.

To estimate body motion, we compute the pixel difference $D_i = (d^x, d^y, d^z)$ between two consecutive frames of the upper centroid C_1 in the x, y, z direction. The upper centroid was selected based on the consideration that the torso orientation is the most characteristic reference during the execution of a full-body action [48]. We then estimate body velocity with respect to the sensor as

$$S_i = \left\{ \frac{d^x}{s}, \frac{d^y}{s}, \frac{d^z}{s} \right\}, \tag{1}$$

where $s = \sqrt{(d^x)^2 + (d^y)^2 + (d^z)^2}$.

We encode S_i as horizontal and vertical speed with respect to the image plane, respectively expressed as $h_i = \sqrt{(S_i^x)^2 + (S_i^z)^2}$ and $v_i = S_i^y$. The former refers to the target moving on the width and depth axis, i.e. closer, further, right, and left. The latter represents the speed with respect to height, e.g. negative if the target is moving down.

For each processed frame i, we obtain the following pose-motion vector:

$$F_i = \left(\theta_i^u, \theta_i^l, h_i, v_i \right). \tag{2}$$

This representation describes spatio-temporal properties of actions in terms of length-invariant, sequential vectors, particularly suitable for serving as input for neural network architectures.

3.2 Learning Framework

Unsupervised neural network learning has shown to be a prominent approach for the detection of abnormal events [49], also referred to as anomaly detection [50]. We propose a hybrid neural-statistical framework to approximate the normal behavior with trained self-organizing map (SOM) networks and subsequently detect behavioral patterns that do not conform to the expected learned behavior with an abnormality test.

The SOM is a competitive neural network introduced by Kohonen [51] that has shown to be a compelling approach for clustering motion expressed in terms of multi-dimensional flow vectors [52–55]. The proposed learning framework consists of three SOM networks. A first network Φ_0 is trained to detect outlier values from the extracted pose-motion vectors caused by tracking errors and sensor noise. After this initial learning phase, the pose-motion vectors are processed again to perform a threshold-based test and remove outliers from the training set. The denoised training set is then fed to a hierarchical SOM-based architecture composed of two networks, Φ_1 and Φ_2, for clustering the subspace of normal actions taking into account spatio-temporal relationships of action sequences. A flow chart of this learning stage is illustrated by Fig. 2.

At detection time, extracted vectors will be denoised and processed through the hierarchy of trained SOM networks. New observations that deviate from the learned

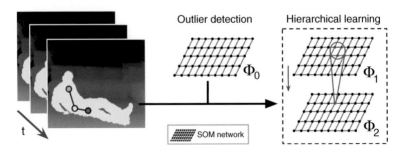

Fig. 2 Flow chart of our SOM-based learning stage. A first network Φ_0 is trained to detect and remove outliers from extracted pose-motion vectors. Preprocessed vectors are fed to a hierarchy of networks (Φ_1 and Φ_2) to cluster spatio-temporal relationships of action sequences

behavior, i.e. below an abnormality threshold, will be reported as abnormal. The detection of noise and abnormal behavior is based on the same abnormality test using two different automatically computed thresholds.

3.2.1 Training Algorithm

The traditional SOM is unsupervised and allows to obtain a low-dimensional discretized representation from high-dimensional input spaces. It consists of a layer with competitive neurons connected to adjacent units by a neighborhood relation. The network learns by iteratively reading each training vector and organizes the units so that they describe the domain space of input observations. Each unit j is associated with a d-dimensional model vector $m_j = [m_{j,1}, m_{j,2}, \ldots, m_{j,d}]$. For each input vector $x_i = (x_1, \ldots, x_n)$ presented to the network, the best matching unit (BMU) b for x_i is selected by the smallest Euclidean distance as

$$b(x_i) = \arg \min_j \|x_i - m_j\|. \tag{3}$$

For an input vector x_i, the quantization error q_i is defined as the distance of x_i from the BMU $b(x_i)$.

We consider two-dimensional networks with units arranged on a hexagonal lattice in the Euclidean space. Each competitive network is trained with a batch variant of the SOM algorithm. This iterative algorithm presents the whole data set to the network before any adjustments are made. The updating is done by replacing the model vector m_j with a weighted average over the samples:

$$m_j(t+1) = \frac{\sum_{i=1}^n h_{j,b(i)}(t) x_i}{\sum_{i=1}^n h_{j,b(i)}(t)}, \tag{4}$$

where b is the best matching unit (Eq. 3), n is the number of sample vectors, and $h_{j,b(i)}$ is a Gaussian neighborhood function:

$$h_{b,i}(x) = \exp\left(\frac{-\|r_b - r_i\|^2}{2\sigma^2(t)}\right), \tag{5}$$

where r_b is the location of b on the map grid and $\sigma(t)$ is the neighborhood radius at time t.

At the second learning stage step, a hierarchical SOM-based approach is used to learn spatio-temporal properties of action sequences from denoised training samples. We first train the network Φ_1 with pose-motion vectors (Eq. 2) from the denoised training set. After this training phase, chains of activated best matching units (Eq. 3) for ordered training sequences produce time varying trajectories on the network map. We empirically define a BMU trajectory for a training vector x_i as

$$\tau_i = (b(x_{i-2}), b(x_{i-1}), b(x_i)). \tag{6}$$

We denote the set of all activation trajectories X as $T(X)$. This step produces a time-selective mapping with action segments from 3 consecutive vectors.

3.2.2 Tracking Errors

An outlier can be seen as an observation that does not follow the pattern suggested by the majority of the observations belonging to the same data cloud [53]. From a geometrical perspective, outliers are to be found detached from the dominating distribution of the subspace of normal actions.

In our approach, we differentiate between outliers introduced by tracking errors and outliers caused by tracked abnormal events. For this purpose, we assume that the behavior of a moving target must be consistent over time. Therefore, we consider highly inconsistent changes in body posture and speed to be caused by tracking errors rather than actual tracked motion. As shown by our experiments, the presence of tracking errors in the training set may negatively affect the SOM-based clustering of pose-motion features. Figure 3 illustrates these effects after the learning phase. A first SOM was trained with the full set of extracted motion vectors, for which outliers in the data decreased the unfolding of the projected feature map (Fig. 3a). These noisy samples were detected by our algorithm and removed from the training set. As seen from the second SOM trained with the denoised training set (Fig. 3b), the absence of outliers allowed a more representative clustering of the motion vectors for the subspace of normal actions.

While we use the same algorithm to detect outliers, two different abnormality thresholds are automatically computed that take into account the different

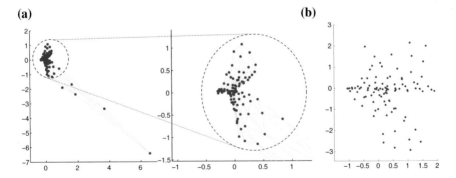

Fig. 3 Effects of outliers in the clustering of training data [9]. **a** The first SOM was trained with the full set of extracted motion vectors. The presence of highly noisy observations in the training set decreased the unfolding of the projected feature map. **b** This second SOM was trained after removing outliers from the training set, resulting in a more representative clustering of the observations from tracked motion

characteristics of tracking noise and abnormal pose-motion vectors. Using this first trained SOM network as reference, also tracking errors in the test set are detected and removed.

3.2.3 Abnormality Detection Algorithm

The goal of the detection algorithm is to test if the most recent observation is abnormal or not. For this purpose, the degree of abnormality for every test observation is expressed with the estimation of a P-value. If the P-value is smaller than a given threshold, then the observation is considered to be abnormal and reported as such.

For a given training set X and a new test observation x_{n+1} presented to the network Φ, the algorithm is summarized as follows [56]:

0. Compute the set of quantization errors $Q = (q_1, q_2, \ldots, q_n)$.
1. Compute $q_{(n+1)}$ with respect to Φ.
2. Define B as the number of quantization errors $(q_1, \ldots, q_n)$ greater than $q_{(n+1)}$.
3. Define the abnormality P-value as $P_{(n+1)} = B/n$.

As an extension of the algorithm proposed in [56], abnormality thresholds are automatically computed for the trained networks Φ_0 and Φ_2. The choice of convenient threshold values that take into account the characteristics of the distributions can have a significant impact on the successful rates for abnormality detection. From a neural network perspective, the threshold values will consider the distribution of the quantization errors from each trained SOM. Based on related research [9], we empirically define two different thresholds, T_O for outlier detection and T_A for abnormality detection:

$$T_O = \beta \sqrt{\overline{Q_o} + \sigma(Q_o)} + \max(Q_o) + \min(Q_o), \tag{7}$$

$$T_A = \gamma \left[\frac{\overline{Q} + \sigma(Q)}{\max(Q) + \min(Q)} \right], \tag{8}$$

where Q_0 and Q denote the quantization error sets for Φ_0 and Φ_2 respectively, $\overline{Q}$ denotes the mean value operator, $\sigma(Q)$ denotes the standard deviation, and $\beta = 0.5$, $\gamma = 0.1$. In the case of Φ_0, observations with P-values under the abnormality threshold T_O are considered as outlier values and therefore removed from the training set. For Φ_2, if $P_{(n+1)}$ is smaller than T_A, the test observation x_{n+1} is considered abnormal.

3.3 Experimental Results

For the acquisition of training data, we monitored a home-like environment with an ASUS Xtion Live sensor installed on a platform 1.30 m above the ground and positioned parallel to the horizontal surface. Depth maps were acquired with a VGA resolution of 640 × 480 and the depth operation range was set from 0.8 to 4 m. The main technical characteristics of the Xtion live sensor are listed in Table 1 [7]. Video sequences were sampled at a constant frame rate of 30 Hz. To reduce sensor noise, we sampled the median value of the last 3 estimated points. Body centroids were estimated from depth map sequences based on the tracking skeleton model provided by the publicly available OpenNI/NITE framework.[1]

For the training phase and the system evaluation, we used video sequences from our data set with full-body actions performed by 13 different participants of the study with a normal physical condition [47]. To avoid biased execution, the participants had not been explained how to perform the actions. Training video sequences consisted of domestic actions such as walking, sitting down, standing up, and bending to pick up objects; abnormal actions consisted in falling down and crawling. We did not take into account those cases in which the user has already fallen on the ground since the tracking framework built on top of OpenNI would fail to provide a reliable recognition of the user and therefore, the extraction of body features would be highly compromised.

At detection time, new extracted vectors were processed to remove outliers. For the last three denoised vectors, a new test trajectory τ_{i+1} was obtained from Φ_1 and then fed to Φ_2 to compute the abnormality test $\lambda(\tau_{i+1})$. We took the last 3 abnormality test results and returned as abnormality output the result of the statistical mode:

$$Mo(\lambda(\tau_{i+1}), \lambda(\tau_{i+2}), \lambda(\tau_{i+3})). \tag{9}$$

A new output was therefore returned every 9 samples, which corresponds to approximately less than 1 s of captured motion. As shown by our experiments, this approach led to increased detection accuracy.

We evaluated the detection algorithm on abnormal actions using standard measurements defined by Van Rijsbergen [57]:

$$Recall = \frac{TP}{TP + FN}, \tag{10}$$

$$Precision = \frac{TP}{TP + FP}, \tag{11}$$

---

[1]OpenNI/NITE: http://www.openni.org/software.

Table 2 Evaluation of our abnormality detection algorithm on a data set of 13 participants

	Raw (%)	Denoised (%)	Improvement (%)
Recall	88	**95**	7.02
Precision	90	**97**	7.02
F-score	89	**96**	7.02
TN rate	90	**97**	6.90
Accuracy	89	**96**	6.96

Best results in bold

$$\text{F-score} = 2 * \frac{\text{Recall} \cdot \text{Precision}}{\text{Recall} + \text{Precision}}, \tag{12}$$

$$\text{True negative rate} = \frac{\text{TN}}{\text{TN} + \text{FP}}, \tag{13}$$

$$\text{Accuracy} = \frac{\text{TP} + \text{TN}}{\text{TP} + \text{TN} + \text{FP} + \text{FN}}. \tag{14}$$

A true positive (TP) was obtained when an abnormal event was detected between the first and the last frame where the abnormal action took place. True negatives (TN) refer to normal actions not detected as abnormal. False positives (FP) and false negatives (FN) refer respectively to normal actions reported as abnormal and abnormal behaviors not reported by the system.

The system evaluation is shown in Table 2. Our system detected abnormal fall and crawling events with 96 % accuracy. The removal of noise from the training and test set was of significant importance for reducing detection errors in presence of partial occlusions and tracking errors introduced by the mobile sensor, with an improvement in accuracy of 6.96 %. On the other hand, the accuracy of our system would be negatively influenced by: (1) highly-occluded users, leading to tracking errors and compromised feature extraction; and (2) the presence of actions sharing similar body features subject to classification ambiguity, i.e. detecting lying down as a fall, leading to a greater number of false positives.

4 Neurocognitive Robot Assistant

We now present a robot assistant that monitors a person in a household environment and reports abnormal user behavior such as a fall event. The underlying idea is that the robot will track a person in the scene while they perform daily activities, thereby tracking the user's position, body posture and speed. The information processed by the tracking framework is fed into the neural system which is responsible for triggering alarms of abnormal events. Whenever an abnormal action, e.g., a fall, is detected, the humanoid will approach the user and ask whether assistance is required. The robot will then take a picture of the scene that can be sent to the user's caregiver for telematic evaluation and agile intervention.

In this section, we first depict an overview of our system and its components. We then go into detail about the software, hardware, and the communication interface, and finally present the experimental set-up for fall detection in a home-like environment.

4.1 System Overview

Our fall detection system consists of a humanoid robot Nao extended with a depth sensor, a tracking framework to keep the user in the scene, and a learning-based system to process the visual information and detect abnormal user behaviors (Fig. 4). All these components communicate over a middleware layer based on Robot Operating System (ROS) that supports different hardware elements and programming languages.

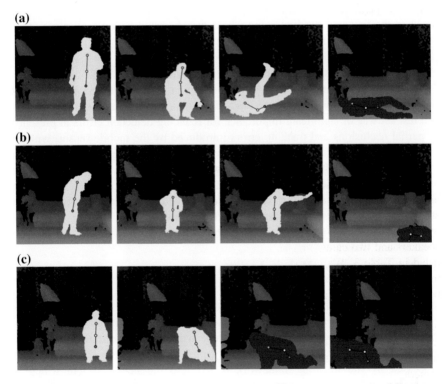

Fig. 4 Abnormal event detection from video sequences. The system can successfully detect abnormal actions and report them (*red body*). **a** Fall event, **b** fall event with partially occluded person, and **c** crawling sequence

Fig. 5 Humanoid Nao
extended with ASUS Xtion
Live sensor on the head [9].
This approach allows to use
Nao's actuators and sensed
depth information to actively
track a moving person in the
environment

Table 3 Nao next gen
specifications

Height	58 cm
Weight	4.3 kg
Autonomy	60–90 min (active/normal use)
Degrees of freedom	21–25
CPU	Intel Atom @ 1.6 GHz
Compatible OS	Linux, Windows, Mac OS
Vision	2 × HD cameras (1280 × 960)
Connectivity	Ethernet, Wi-Fi

Robot Nao (shown in Fig. 5) is a middle-size mobile humanoid robot developed by Aldebaran Robotics.[2] Since the beginning of the Nao project in 2004, the humanoid underwent a significant number of enhancements, thereby becoming the standard robot platform for a number of research institutions and robot competitions (e.g. RoboCup[3]). Nao includes an embedded multimedia system with microphones, speakers and two cameras. The main technical characteristics of Nao Next Gen are listed in Table 3. We extended the robot Nao with an ASUS Xtion depth sensor installed on top of the head (Fig. 5). The Xtion sensor was chosen over the Kinect because of its reduced power consumption and weight. A set of experiments with the extended Nao showed that wearing the sensor does not affect the overall stability of the humanoid while standing or walking. In contrast to the use of a fixed sensor, Nao will pan its head to seamlessly keep the moving person in the scene. In case of a fall event detection, a color picture of the scene will be taken using Nao's camera.

---

[2]Aldebaran Robotics: http://www.aldebaran-robotics.com/.

[3]RoboCup Project: http://www.robocup.org/.

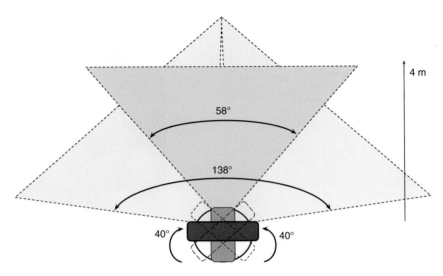

Fig. 6 Nao with Xtion sensor: extended horizontal field of view from 58° to 138° with a maximum head pan angle of 40° in each direction

4.2 Tracking with a Mobile Sensor

Depth sensors such as Microsoft Kinect and ASUS Xtion Live Pro are characterized by a reduced field of view (58 horizontal, 45 vertical, 70 diagonal), and therefore limiting their use in expansive environments. The idea of active tracking consists of seamlessly keeping the person in the scene, thereby moving the sensor when the person is approaching an area outside the field of view (FOV).

We use Nao's head to move the sensor and increase the horizontal FOV from 58° to 138° (Fig. 6). As a strategy for active tracking, we define a bounding box in which the target can act without the sensor being moved (Fig. 7a). We consider the upper-body centroid as the reference of the person's position. When the centroid lies outside the threshold, the tracking application will compute the needed operations to keep the person within the bounding box (Fig. 7b). Nao will then smoothly pan its head by 10° in the required direction, for a maximum pan angle of 40° in each direction (Fig. 6).

The body tracking application is built on top of simple-openni [58], which wraps the OpenNI–NITE framework for user identification, calibration and estimation of skeletal joints. We use this library with Processing IDE[4] with the purpose to enable a simplified access to some functionalities provided by the OpenNI such as skeleton tracking and scene analysis.

All system modules for active tracking communicate over Robot Operating System (ROS), a software framework for robot software development with

---

[4]Processing IDE: http://processing.org/.

(a)   **(b)**

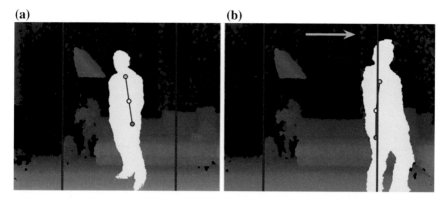

Fig. 7 Threshold-based active tracking strategy. When the upper-body centroid lies outside the threshold, the tracking application will compute the needed operations to keep the person within the bounding box (*red lines*)

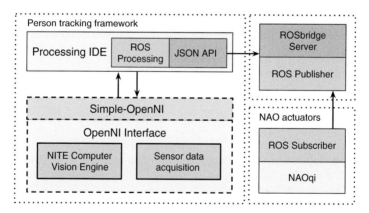

Fig. 8 A diagram of the communication network for interfacing the tracking framework with Nao's actuators over ROS

operating system-like functionality on a heterogeneous computer cluster [59]. It provides hardware abstraction, device drivers, libraries, visualizers, message-passing between processes and package management. A diagram of the overall architecture for active tracking is illustrated in Fig. 8. To interface our system modules, we use a ROS-based communication network implemented with *publisher-subscriber* nodes. We implement publisher nodes to continually broadcast a message over the network using a message-adapted class. The subscriber node will receive the messages on a given topic via a master node, which keeps a registry of publishers and subscribers. This specific architecture represents a robust interface to connect different applications, e.g. written in different programming languages, over a common network of communication. The tracking framework

Fig. 9 Person monitoring in a home-like environment. The Nao will seamlessly track the person while performing daily activities

communicates to ROS over Rosbridge[5] and a modified version of ROSProcessing,[6] extended to publish ROS topics. Rosbridge provides a JSON API[7] to ROS functionality for non-ROS programs. The rosbridge_suite package is a collection of packages that implement the rosbridge protocol and provides a WebSocket transport layer. We program Nao to move its head according to the tracking application via NAOqi framework,[8] which allows homogeneous communication between different Nao modules (motion, audio, video), and ROS integration.

4.3 Fall Detection Scenario

To test our system, we run the experiments in a home-like environment with a person performing daily activities, such as walking around the room, bending to collect objects, and sitting down to read (Fig. 9). The Nao was initially positioned on one side of the room to monitor the scene and connected to the system using wireless communication. The depth sensor was connected to a laptop (i5-3320 M

---

[5]ROSbridge_suite: http://wiki.ros.org/rosbridge_suite.

[6]ROSProcessing: https://github.com/pronobis/ROSProcessing.

[7]JSON API: http://jsonapi.org/.

[8]NAOqi framework: https://community.aldebaran-robotics.com/doc/1-14/dev/naoqi/index.html.

(a)   (b)

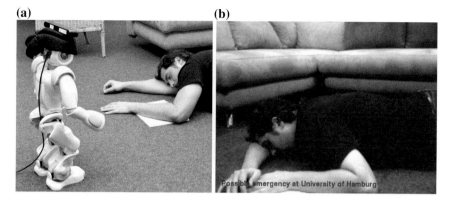

Fig. 10 Fall detection scenario: In case a fall event is detected by the system, the Nao will approach the fallen person (**a**) and take a picture of the scene (**b**)

2.6 GHz Processor and 4 GB of RAM) running all system modules under Ubuntu desktop 12.04[9] and ROS Groovy.[10] Whenever the person approaches the edge of the field of view of the sensor, Nao will pan its head to keep him/her in the scene. When a fall event is detected by the neural framework, the system will trigger an alarm. As shown in Fig. 10, the humanoid will approach the person by using the last tracked position before the fall and ask whether assistance is required. The color camera will be used to take a picture of the fallen person that can be sent to a relative or to the person's caregiver for further human assessment.

5 Discussion

The robust detection of falls in home environments represents a paramount component for assistive systems aiming to enhance the person's safety perception and avoid the loss of confidence due to, for instance, functional disabilities. In this context, vision-based fall detection has been shown to be a predominant approach due to the substantial advances in computer vision techniques and reduced cost factors with respect to wearable sensors. In this setting, the visual recognition of human actions is a key issue introducing a vast set of challenges for traditional 2D cameras. The use of low-cost depth sensors capable of performing 3D human motion segmentation and body posture estimation has led to promising results in experimental scenarios. However, despite the latest sensor trends, the question remains open on how to better process extracted body features for effectively extrapolating the complex dynamics of actions and fall events, exhibiting noise

---

[9]Ubuntu Desktop: http://www.ubuntu.com/desktop.

[10]ROS Groovy Galapagos: http://wiki.ros.org/groovy.

tolerance and robustness in real-world scenarios. Learning-based paradigms such as machine learning and neural networks have been shown to be a promising methodology for achieving knowledge generalization of a set of training activities and the subsequent classification of unseen situations [8, 9, 47].

In Sect. 3, we presented research on abnormal event detection based on unsupervised neural network learning. Our system consists of a hierarchical neural architecture that can learn a set of normal actions, e.g. walking, sitting, and lying down, captured by a depth sensor. After the training phase, the system will report novel behavioral patterns, e.g. fall events, as abnormal actions and trigger an alarm. The combination of a depth sensor with our neural network approach allows to tailor the robust detection of fall events independently from the background surroundings and changing light conditions. In addition, to contrast sensor noise and tracking errors, the neural architecture is also responsible for automatically removing noisy samples from the extracted body features during the training and test stage. Experiments run in a home-like environment showed that our system can detect fall events with high accuracy in real time (as shown in Table 2).

Contrary to the use of fixed ambient sensors, mobile assistive robots can undertake actions that benefit people with disabilities and seniors in a residential context. As supported by recent studies, socially-aware assistive solutions can provide positive effects on the senior's well-being in domestic environments [10], for instance, by supporting basic daily tasks of independent living and enhancing the user's experience through flexible human-robot interaction. On the other hand, this technology introduces new technical challenges and issues.

In Sect. 4, we introduced a humanoid robot to assist a person in a household environment and report abnormal user behavior such as fall events. The underlying motivation is to use a mobile robot to track the user's position, body posture, and motion characteristics when the user is performing daily activities. The processed visual information from the mobile depth sensor is fed into our neural system for abnormality detection. The removal of noise is of significant importance for reducing detection errors in presence of partial occlusions and tracking errors introduced by the mobile sensor, with an improvement in detection accuracy of 6.96 %. In case a fall event is detected, the humanoid will approach the user and then take a picture of the scene that can be sent to the user's caregiver for telematic evaluation and agile intervention. For our experiments, we did not consider those cases in which the user has already fallen on the ground when the robot starts to monitor the scene. This is due to the fact that a fallen person would not be detected by the tracking framework built on top of OpenNI that works better with moving users for user calibration and pose estimation. Therefore, the reliable detection and segmentation of a fallen user are open issues to be addressed, e.g. by using complementary RGB information to recognize a body on the ground [60].

The obtained results motivate future work in several directions. For instance, the ability of the robot to navigate in the environment for following the person through different rooms and finding a better angle of view to avoid body occlusions. At the current state of the system, the depth sensor must be wired to an external, fixed processing unit to perform the tracking, thereby limiting the mobility of the humanoid. To achieve better mobility, the sensor could be wired to an onboard processing unit and then transmit the depth information via WiFi for further processing to be carried out in the cloud. Moreover, video files could be adopted instead of a single picture to better support telematic human evaluation, e.g. sending a video with the last five seconds of the user's activity before the fall event. In fact, the role of human assessment is of crucial importance to determine the seriousness of the detected event and to undertake effective intervention.

To cope with the dynamic nature of real-world scenarios, a learning artificial system may not only be robust to unseen situations, but also adaptive. In fact, in addition to detecting short-term behavior such as fall events and domestic daily actions (e.g. walking, drinking, lying down), it may be of particular interest to monitor and learn the user's behavior over longer periods of time [5]. In this setting, it would be desirable to collect sensory data to, e.g., perform medium- and long-term gait assessment of the person, which can be an important indicator for a variety of health problems, e.g. physical diseases and neurological disorders such as Parkinson's disease [61]. To enhance the user's experience, assistive robots may be given the capability to adapt over time to better interact with the monitored user. This would include, for instance, a more natural human-robot communication including the recognition of hand gestures and full-body actions, speech recognition, and a set of reactive behaviors based on the user's habits. In this context, interdisciplinary research that takes into account the vast set of technical, social, and ethical issues regarding robots for assisted living is fundamental to provide feasible and reliable solutions in the near future.

Acknowledgements The authors would like to thank Erik Strahl for his invaluable technical contribution and help. The authors gratefully acknowledge funding by the DAAD German Academic Exchange Service (Kz:A/13/94748)—Cognitive Assistive Systems Project, by the DFG German Research Foundation (grant #1247)—International Research Training Group CINACS (Cross-modal Interaction in Natural and Artificial Cognitive Systems), and the DFG under project CML (TRR169).

References

1. World Health Organization: Global Report on Falls Prevention in Older Age. http://www.who.int/ageing/publications/Falls_prevention7March.pdf
2. Tinetti, M.E., Liu, W.L., Claus, E.B.: Predictors and prognosis of inability to get up after falls among elderly persons. J. Am. Med. Assoc. **269**(1), 65–70 (1993)
3. Scheffer, A.C., Schuurmans, M.J., van Dijk, N., van der Hooft, T., de Rooij, S.E.: Fear of falling: measurement strategy, prevalence, risk factors and consequences among older persons. Age Ageing **37**(1), 19–24 (2008)
4. Kaluza, B., Cvetkovic, B., Dovgan, E., Gjoreski, H., Gams, M., Lustrek, M.: A multi-agent care system to support independent living. Int. J. Artif. Intell. Tools **23**(1), 1–20 (2013)
5. Vettier, B., Garbay, C.: Abductive agents for human activity monitoring. Int. J. Artif. Intell. Tools **23** (2014)
6. Microsoft Kinect for Windows: http://www.microsoft.com/en-us/kinectforwindows/. Cited 10 Sept 2014

7. ASUS Xtion PRO LIVE sensor: http://www.asus.com/Commercial_3D_Sensor/Xtion_PRO_LIVE. Cited 10 Sept 2014
8. Jiang, Z., Lin, Z., Davis, L.S.: Recognizing human actions by learning and matching shape-motion prototype trees. IEEE Trans. Pattern Anal. Mach. Intell. **31**(3), 533–547 (2012)
9. Parisi, G. I., Wermter, S.: Hierarchical som-based detection of novel behavior for 3D human tracking. In: Proceedings of the IEEE International Joint Conference on Neural Networks (IJCNN), pp. 1380–1387, Dallas, Texas, USA (2013)
10. Kachouie, R., Sedighadeli, S., Khosla, R., Chu, M.-T.: Socially assistive robots in elderly care: a mixed-method systematic literature review. Int. J. Hum. Comput. Interact. **30**(5), 369–393 (2014)
11. Igual, R., Medrano, C., Plaza, I.: Challenges, issues and trends in fall detection systems. BioMed. Eng. OnLine 12–66 (2013)
12. Mubashir, M., Shao, L., Seed, L.: A survey on fall detection: principles and approaches. Neurocomputing **100**, 144–152 (2013)
13. Bourke, A.K., de Ven, V., Gamble, M., O'Connor, R., Murphy, K., Bogan, E., McQuade, E., Finucane, P., OLaighin, G., Nelson, J.: Assessment of waist-worn tri-axial accelerometer based fall-detection algorithms using continuous unsupervised activities. In: Proceedings of the Annual International Conference of the IEEE Engineering in Medicine and Biology Society, pp. 2782–2785. Institute of Electrical and Electronics Engineers, Buenos Aires (2010)
14. Lai, C.F., Chang, S.Y., Chao, H.C., Huang, Y.M.: Detection of cognitive injured body region using multiple triaxial accelerometers for elderly falling. IEEE Sens. J. **11**, 763–770 (2011)
15. Kerdegari, H., Samsudin, K., Ramli, A.R., Mokaram, S.: Evaluation of fall detection classification approaches. In: Proceedings of the 4th International Conference on Intelligent and Advanced Systems, pp. 131–136. Institute of Electrical and Electronics Engineers, Kuala Lumpur (2012)
16. Cheng, J., Chen, X., Shen, M.: A framework for daily activity monitoring and fall detection based on surface electromyography and accelerometer signals. IEEE J. Biomed. Health Inf. **17**(1), 38–45 (2013)
17. Albert, M.V., Kording, K., Herrmann, M., Jayaraman, A.: Fall classification by machine learning using mobile phones. PLoS ONE **7**, e36556 (2012)
18. Lee, R.Y.W., Carlisle, A.J.: Detection of falls using accelerometers and mobile phone technology. Age Ageing **0**, 1–7 (2011)
19. Fang, S.H., Liang, Y.C., Chiu, K.M.: Developing a mobile phone-based fall detection system on android platform. In: Proceedings of the Conference on Computing, Communications and Applications, pp. 143–146, Hong Kong (2012)
20. Abbate, S., Avvenuti, M., Bonatesta, F., Cola, G., Corsini, P., Vecchio, A.: A smartphone-based fall detection system. Pervasive Mob. Comput. **8**, 883–899 (2012)
21. Patsadu, O., Nukoolkit, C., Watanapa, B.: Survey of smart technologies for fall motion detection: techniques, algorithms and tools. In: Papasratorn, B., et al. (eds.) IAIT 2012, CCIS 344, pp. 137–147. Springer, Heidelberg (2012)
22. Botia, J.A., Villa, A., Palma, J.: Ambient assisted living system for in-home monitoring of healthy independent elders. Expert Syst. Appl. **39**, 8136–8148 (2012)
23. Lee, T., Mihailidis, A.: An intelligent emergency response system: preliminary development and testing of automated fall detection. J. Telemed. Telecare **11**, 194–198 (2005)
24. Miaou, S.G., Sung, P.H., Huang, C.Y.: A customized human fall detection system using omni-camera images and personal information. In: Proceedings of the 1st Distributed Diagnosis and Home Healthcare Conference, pp. 39–42. Institute of Electrical and Electronics Engineers, Arlington (2006)
25. Rougier, C., Meunier, J., St-Arnaud, A., Rousseau, J.: Robust video surveillance for fall detection based on human shape deformation. IEEE Trans. Circuits Syst. Video Technol. **21**, 611–622 (2011)
26. Liu, C.L., Lee, C.H., Lin, P.M.: A fall detection system using k-nearest neighbor classifier. Expert Syst. Appl. **37**, 7174–7181 (2010)

27. Cucchiara, R., Prati, A., Vezzani, R.: A multi-camera vision system for fall detection and alarm generation. Expert Syst. **24**, 334–345 (2007)
28. Hazelhoff, L., Han, J., de With, P.H.N.: Video-based fall detection in the home using principal component analysis. In: Bland-Talon, J., Bourennane, S., Philips, W., Popescu, D., Scheunders, P. (eds.) Proceedings of the 10th International Conference on Advanced Concepts for Intelligent Vision Systems, pp. 298–309. Springer, Juan-les-Pins (2008)
29. Diraco, G., Leone, A., Siciliano, P.: An active vision system for fall detection and posture recognition in elderly healthcare. In: Conference and Exhibition: Design, Automation and Test in Europe, pp. 1536–1541. European Design and Automation Association, Dresden (2010)
30. Han, J., Shao, L., Xu, D., Shotton, J.: Enhanced computer vision with microsoft kinect sensor: a review. IEEE Trans. cybern. **43**(5), 1318–1334 (2013)
31. Rougier, C., Auvinet, E., Rousseau, J., Mignotte, M., Meunier, J.: Fall detection from depth map video sequences. In: Abdulrazak, B., et al. (eds.) ICOST 2011. LNCS 6719, pp. 121–128 (2011)
32. Planinc, R., Kampel, M.: Introducing the use of depth data for fall detection. In: Personal Ubiquitous Computing, vol. 17, pp. 1063–1072. Springer, Heidelberg (2012)
33. Mastorakis, G., Makris, D.: Fall detection system using Kinects infrared sensor. J. Real-Time Image Process (2012) (Springer)
34. Zhang, C., Tian, Y., Capezuti, E.: Privacy preserving automatic fall detection for elderly using RGBD cameras. In: Miesenberger, K., Karshmer, A., Penaz, P., Zagler, W. (eds.) Proceedings of the 13th International Conference on Computers Helping People with Special Needs, pp. 625–633. Springer, Linz (2012)
35. Gasparrini, S., Cippitelli, E., Spinsante, S., Gambi, E.: A depth-based fall detection system using a kinect sensor. Sensors **14**, 2756–2775 (2014)
36. Cogniron: Cognitive Robot Companion. http://www.cogniron.org. Cited 15 Jan 2015
37. LIREC: Living with Robots and Interactive Companions. http://lirec.eu/. Cited 15 Jan 2015
38. Hermes: Cognitive Care and Guidance for Active Ageing. http://www.fp7-hermes.eu. Cited 15 Jan 2015
39. KSERA: Knowledgable SErvice Robots for Aging. http://ksera.ieis.tue.nl. Cited 15 Jan 2015
40. GiraffPlus: http://www.giraffplus.eu. Cited 15 Jan 2015
41. ROBOT-ERA: Implementation and integration of advanced robotic systems and intelligent Environments in real scenarios for the ageing population. http://www.robot-era.eu. Cited 15 Jan 2015
42. Amirabdollahian, F., Bedaf, S., Bormann, R., Draper, H., Evers, V., Gallego Perez, J., Gelderblom, G.J., et al.: Assistive technology design and development for acceptable robotics companions for ageing years. Paladyn J. Behav. Robot. **4**(2), 1–9 (2013)
43. Mundher, Z.A., Zhong, J.: A real-time fall detection system in elderly care using mobile robot and kinect sensor. Int. J. Mater. Mech. Manuf. **2**(2), 133–138 (2014)
44. Volkhardt, M., Schneemann, F., Gross, H.-M.: Fallen person detection for mobile robots using 3D depth data. In: Proceedings of IEEE International Conference on Systems, Man, and Cybernetics (IEEE-SMC), pp. 3573–3578, Manchester, GB (2013)
45. Martinson, E.: Detecting occluded people for robotic guidance. In: IEEE International Symposium on Robot and Human Interactive Communication (RO-MAN), pp. 744–749, Edinburgh, Scotland, UK (2014)
46. Volkhardt, M., Gross, H.-M.: Finding people in home environments with a mobile robot. In: European Conference on Mobile Robots (ECMR), pp. 282–287, Barcelona, Spain (2013)
47. Parisi, G.I., Weber, C., Wermter, S.: Human action recognition with hierarchical growing neural gas. In: Wermter, S., et al. (eds.) Proceedings of the International Conference on Artificial Neural Networks (ICANN), pp. 89–96, Hamburg, Germany (2014)
48. Papadopoulos, G.Th., Axenopoulos, A., Daras, P.: Real-time skeleton-tracking-based human action recognition using kinect data. In: Gurrin, C., et al. (eds.) MultiMedia Modeling. LNCS, vol. 8325, pp. 473–483, Springer International Publishing (2014)
49. Hu, W., Tan, T., Wang, L., Maybank, S.: A survey on visual surveillance of object motion and behaviors. IEEE Trans. Syst. Man Cybern. **34**(3), 334–352 (2004)

50. Chandola, V., Banerjee, A., Kumar, V.: Anomaly detection: a survey. ACM Comput. Surveill. **15** (2009)
51. Kohonen, T.: Self-organizing map, 2nd edn. Springer, Heidelberg (1995)
52. Hu, W., Xie, D., Tan, T.: A hierarchical self-organizing approach for learning the patterns of motion trajectories. IEEE Trans. Neural Netw. **15**(1), 135–144 (2004)
53. Nag, A.K., Mitra, A., Mitra, S.: Multiple outlier detection in multivariate data using self-organizing maps title. Comput. Stat. **2**(2), 245–264 (2005)
54. Parisi, G.I., Barros, P., Wermter, S.: FINGeR: framework for interactive neural-based gesture recognition. In: Proceedings of the European Symposium on Artificial Neural Networks, Computational Intelligence and Machine Learning (ESANN), pp. 443–447, Bruges, Belgium (2014)
55. Parisi, G.I., Jirak, D., Wermter, S.: HandSOM: neural clustering of hand motion for gesture recognition in real time. In: Proceedings of the IEEE International Symposium on Robot and Human Interactive Communication (RO-MAN), pp. 981–986, Edinburgh, Scotland, UK (2014)
56. Hoglund, A.J., Hatonen, K., Sorvari, A.S.: A computer host-based user anomaly detection system using self-organizing maps. In: IEEE-INNS-ENNS International Joint Conference on Neural Networks, vol. 5, pp. 411–416 (2000)
57. Van Rijsbergen, C.J.: Information retrieval, 2nd edn. Information Retrieval, Butterworth-Heinemann, London (1979)
58. simple-openni—OpenNI library for Processing: https://code.google.com/p/simple-openni/. Cited 15 Jan 2015
59. The Robot Operating System (ROS): http://www.ros.org/. Cited 15 Jan 2015
60. Wang, S., Zabir, S., Leibe, B.: Lying pose recognition for elderly fall detection. In: Conference on Robotics: Science and Systems VII, pp. 44–50, Los Angeles, CA, USA (2011)
61. Aerts, M.B., Esselink, R.A.J., Post, B., van de Warrenburg, B.P.C., Bloem, B.R.: Improving the diagnostic accuracy in parkinsonism: a three-pronged approach. Pract. Neurol. **12**(2), 77–87 (2012)

Authors Biography

German I. Parisi received his Bachelor's and Master's degree in Computer Science from the University of Milano-Bicocca, Italy. Since 2013, he is a research associate in the Knowledge Technology Group at the University of Hamburg, Germany, where he is part of the research project CASY (Cognitive Assistive Systems) and the international Ph.D. research training group CINACS (Cross-Modal Interaction in Natural and Artificial Cognitive Systems).

His main research interests include neurocognitive systems for human-robot assistance, computational models of the visual cortex, and bio-inspired action and gesture recognition.

Stefan Wermter received the Diploma from the University of Dortmund, the M.Sc. from the University of Massachusetts, and the Ph.D. and Habilitation from the University of Hamburg, all in Computer Science. He has been a visiting research scientist at the International Computer Science Institute in Berkeley before leading the Chair in Intelligent Systems at the University of Sunderland, UK. Currently Stefan Wermter is Full Professor in Computer Science at the University of Hamburg and Director of the Knowledge Technology institute.

His main research interests are in the fields of neural networks, hybrid systems, cognitive neuroscience, bio-inspired computing, cognitive robotics and natural language processing. In 2014 he was general chair for the International Conference on Artificial Neural Networks (ICANN). He is also on the current board of the European Neural Network Society, associate editor of the journals "Transactions on Neural Networks and Learning Systems", "Connection Science", "International Journal for Hybrid Intelligent Systems" and "Knowledge and Information Systems" and he is on the editorial board of the journals "Cognitive Systems Research" and "Journal of Computational Intelligence".

Smart Robot Control via Novel Computational Intelligence Methods for Ambient Assisted Living

Bo Xing

Abstract In recent years, we have witnessed a rapid surge in ambient assisted living (AAL) technologies due to a rapidly aging society. The aging population, the increasing cost of formal health care, the caregiver burden, and the importance that the individuals place on living independently, all motivate development of innovative-assisted living technologies for safe and independent aging. Among all areas, the development of assistive robots with potential application to health and elderly care is drawing a lot attention from both industry and academia. Several such prototypes (e.g., wearable assisted-walking device designed by Honda, and personal transport assistance robot designed by Toyota) have been made. These robots are inevitably supported by various algorithms and computational techniques. In this work, we provide an overview of applying emerging CI (i.e., non-conventional CI) approaches to various smart robot control scenarios which, from the author's viewpoint, have a great influence on various ambient assisted living (AAL) robot related activities, such as location identification, manipulation, communication, vision, learning, and docking capabilities. The innovative CI methods covered in this chapter include bacteria foraging optimization (BFO), bees algorithm (BA), glowworm swarm optimization (GSO), grey wolf optimizer (GWO), electromagnetism-like mechanism (EM), intelligent water drops (IWD), and gravitational search algorithm (GSA). The findings of this work can provide a good source for someone who is interested in the research filed of AAL robot.

Keywords Ambient assisted living (AAL) · Robot control · Computational intelligence (CI) · Bacteria foraging optimization (BFO) · Bees algorithm (BA) · Glowworm swarm optimization (GSO) · Electromagnetism-like mechanism (EM) · Intelligent water drops (IWD) · Gravitational search algorithm (GSA) · Grey wolf optimizer (GWO)

B. Xing (✉)
Computational Intelligence, Robotics, and Cybernetics for Leveraging E-future (CIRCLE),
Department of Computer Science, School of Mathematical and Computer Sciences,
Faculty of Science and Agriculture, University of Limpopo,
Private Bag X1106, 0727 Sovenga, Limpopo, South Africa
e-mail: bxing2009@gmail.com

© Springer International Publishing Switzerland 2016
K.K. Ravulakollu et al. (eds.), *Trends in Ambient Intelligent Systems*,
Studies in Computational Intelligence 633, DOI 10.1007/978-3-319-30184-6_2

29

1 Introduction

Technology and automation have made a great number of labour-saving robots become possible. For example, numerous monotonous and dangerous tasks such as sheet metal painting and oil pipelines cleaning used to be completed by humans have been substituted by robots. More recently, the implementation of ambient assisted living (AAL) robots that are capable of assisting human beings, in particular in the scenario of offering assistance to the disabled and the elderly, has become more and more popular (e.g., MOVAID system [1, 2]). However, there are times that these automated devices poorly perform, especially in the situation of when they really needs to interact with another human being [3]. At present, different computational intelligence (CI) approaches such as neural networks [4], reinforcement learning [5–7], and fuzzy logic [8] are helping us to enable a higher level of robot intelligence. Nevertheless, in spite of these algorithms' tremendous success, none of them performs predominant over the rest. To overcome this issue, researchers come up two solutions: one is to use the hybridized version of these algorithms in which the strengths of each algorithm is enhanced while the weaknesses are largely compensated or eliminated (e.g., [9–14]); the other one is to develop new intelligent algorithms which can perform more powerful than their predecessors. Although the first trend is currently well-documented, the second trend is being less mentioned in the literature. The focus of this study is thus placed on the latter trend that is, exploring the novel CI methods and pinpointing their applications in the field of robot control with a particular emphasis, if possible, on AAL robot relevant studies.

With this in mind, the remainder of this chapter is organized as follows: First, Sect. 2 briefly introduces the relevant terms used in this study; Second, the innovative CI algorithms (i.e., biology- and physics-based CI) and their selected applications in smart robot control are detailed in Sects. 3 and 4, respectively; an experimental study that implementing an innovative CI algorithm (i.e., grey wolf optimizer) for the heterogeneous robot swarm control has been described in Sect. 5, which is followed by future work highlighted in Sect. 6; Finally, Sect. 7 draws the conclusion of this study.

2 Background

2.1 What Is Ambient Assisted Living (AAL) Robot?

The main purpose of introducing AAL robots is to assist the disabled and elderly people at home [15–17]. In general, they can be categorized into three categories: robots assisting with daily living activities (such as feeding and dressing), robots assisting with instrumental activities of daily living (such as housekeeping and preparing food), and robots assisting with enhanced activities of daily living (such as communication and engaging in hobbies) [16]. Nowadays, a number of assistive robots are being deployed. For example, "Care-O-bot" [18–20], a robot developed by Fraunhofer IPA, is able to fetch and carry objects, communicate with older people, and supply emergency support; "RIBA" [21, 22], a robot developed by RIKEN-TRI Collaboration Center, can help patient transfer; "uBot5" [23], another robot from the University of Massachusetts Amherst, is capable of achieving multiple postures for the purpose assisting elderly in compensating for impaired upper extremity function; "PerMMA" [24–26], a research outcome from the Carnegie Mellon University and the University of Pittsburgh, can assist persons with disabilities; "PaPeRo" [27–29], another case developed by NEC, is used to communicate; and "Emiew" [30–32] developed by Hitachi, can interact with human beings; and "Hospi-R" [33, 34], an autonomous delivery robot developed by Matsushita, can even perform complex service tasks.

2.2 What Is Robot Control?

As we know, designing assistive living robots is a multi-dimensional, interdisciplinary challenge, which including design challenges, pervasive computing challenges, social challenges, safety challenges, etc. Among these challenges, the control of an autonomous mobile robot is a complex, usually non-linear and partly unpredictable process. The most distinguished difference, between robots and other technologies, lies in that robots' capability to combine automation with action, and in the meantime a considerable amount of mobility. Static tasks only form part of their job can-do list, robots can also do lots of dynamic jobs such as exploration, walking, and lifting. Robots are playing an increasingly important role in both scientific and our daily life. The ultimate goal of the robot control is thus to let robots seamless integrated into a human environment, or in other words, to become more and more humanoid.

2.3 What Is Novel Computational Intelligence?

Computational intelligence (CI) is a fairly new research field, which is still in a process of evolution. At a more general level, CI comprises a set of computing systems with the ability to learn and deal with new events/situations, such that the systems are perceived to have one or more attributes of reason and intelligence [35]. According to their popularity, the CI methods can be roughly classified into two categories, namely, traditional CI and novel CI. Typically, traditional CI approaches include such as simulated annealing, genetic algorithm, particle swarm optimization, ant colony optimization, Tabu search, and artificial neural network. In terms of novel CI approaches [36], this overview covers the following candidates: bacterial foraging optimization (BFO), bees algorithm (BA), glowworm swarm optimization (GSO), electromagnetism-like mechanism (EM), and intelligent water drops (IWD).

3 Biology-Based CI

In this section, we detail a group of biology-based algorithms which get their inspiration from different animals. The representative applications of these algorithms in dealing with various robot control problems are also highlighted.

3.1 Bacteria Foraging Optimization (BFO) Algorithm

Bacterial foraging optimization (BFO) algorithm was originally proposed in Passino [37] where the foraging strategy of E. Coli bacteria has been simulated. Typically, the BFO consists of four main mechanisms: chemotaxis, swarming, reproduction, and elimination-dispersal event. The main steps of BFA are outlined as below [37]:

- Chemotaxis: The movement of E. coli bacteria can be performed through two different ways: swimming which means the movement in the same direction, and tumbling refers the movement in a random direction. The movement of the ith bacterium after one step is given by Eqs. (1) and (2), respectively [37]:

$$\theta^i(j+1,k,l) = \theta^i(j,k,l) + C(i)\phi(j). \qquad (1)$$

$$\begin{cases} \theta^i(j+1,k,l) > \theta^i(j,k,l), & \text{swimming in which } \phi(j) = \phi(j-1) \\ \theta^i(j+1,k,l) < \theta^i(j,k,l), & \text{tumbling in which } \phi(j) \in [0,2\pi] \end{cases} \qquad (2)$$

where $\theta^i(j,k,l)$ denotes the location of ith bacterium at jth chemotactic, kth reproductive and lth elimination and dispersal step, $C(i)$ is the length of unit walk, and $\phi(j)$ is the direction angle of the jth step.

- Swarming: Under the stresses circumstances, the bacteria release attractants to signal bacteria to swarm together, while they also release a repellant to signal others to be at a minimum distance from it. The cell to cell signalling can be represented by Eq. (3) [37]:

$$J_{cc}(\theta, P(j,k,l)) = \sum_{i=1}^{S} J_{cc}^{i}\left(\theta, \theta^{i}(j,k,l)\right)$$

$$= \sum_{i=1}^{S}\left[-d_{attract} \exp\left(-\omega_{attract} \sum_{m=1}^{P} \left(\theta_m - \theta_m^i\right)^2\right)\right]$$

$$+ \sum_{i=1}^{S}\left[h_{repellant} \exp\left(-\omega_{repellant} \sum_{m=1}^{P} \left(\theta_m - \theta_m^i\right)^2\right)\right]. \quad (3)$$

where $J_{cc}(\theta, P(j,k,l))$ is the objective function value to be added to the actual objective function, S is the total number of bacteria, P is the number of variables to be optimized which are present in each bacterium, $\theta = [\theta_1, \theta_2, \ldots, \theta_P]^T$ denotes a point in the P-dimensional search domain, $d_{attract}$ is the depth of the attractant released by the cell, $\omega_{attract}$ is a measure of the width of the attractant signal, $h_{repellant}$ is the height of the repellant effect (i.e., $h_{repellant} = d_{attract}$), and $\omega_{repellant}$ is the measure of the width of the repellant.

- Reproduction: After N_c chemotaxis steps, the reproduction step should be performed. The fitness value of the bacteria is stored in an ascending order. The working principle is the least health bacteria eventually die and the remaining bacteria (i.e., healthiest bacteria) will be divided into two identical ones and placed at the same location.
- Dispersion and elimination: For the purpose to avoid local optima, dispersion and elimination process is performed after a certain number of reproduction steps. According to a present probability (p_{ed}), a bacterium is chosen to be dispersed and moved to another position within the environment.

Taking into account the key phases described above, the steps of implementing BFO can be summarized as follows [37–40]:

- Step 1: Defining the optimization problem, and initializing the optimization parameters.
- Step 2: Iterative algorithm for optimization. In this step the bacterial population, chemotaxis loop $(j = j + 1)$, reproduction loop $(k = k + 1)$, and elimination and dispersal operations loop $(l = l + 1)$ are performed.
- Step 3: If $j < N_c$, go to the chemotaxis process.
- Step 4: Reproduction.
- Step 5: If $k < N_{re}$, go to reproduction process.
- Step 6: Elimination-dispersal.
- Step 7: If $l < N_{ed}$, then go to elimination-dispersal process; otherwise end.

Smart Robot Control using BFO. Based on the basic BFO, an improved version was proposed in [41] to deal with robot navigation problem. Two simulation scenarios, namely, (i) fixed obstacle and target; and (ii) randomly moving obstacles and fixed target, were considered by the authors of [41]. 10 standard deviation values of path lengths were assigned to each testing scenarios where the obtained results are classified into three groups, i.e., best, worst, and average. By comparison the experiment results, both BFO algorithms (improved and basic version) outperform the traditional particle swarm optimization algorithm. At the end of their study, the authors claimed that BFO is very powerful in fulfilling the robot path navigation task.

3.2 Bees Algorithm (BA)

Inspired by natural foraging behavior of honey bees, the bees algorithm (BA) was proposed by the authors of [42]. The basic working principles of BA are listed as follows:

- Step 1: Initializing population with random solutions. At this stage, the BA requires several parameters to be set such as number of scout bees (n), number of sites selected out of n visited sites (m), number of the best sites out of m selected sites (e), number of bees recruited for the best e sites (nep), number of bees recruited for the other $(m - e)$ selected sites (nsp), and initial size of patches (ngh) which includes the site, its neighborhood, and the stopping criterion.
- Step 2: Assessing the fitness of the population.
- Step 3: Forming new population while stopping criterion is not met.
- Step 4: Selecting sites for neighborhood search. The bees with the highest fitness values are chosen as "selected bees" at Step 4, and accordingly, the sites visited by them are chosen for neighborhood search.
- Step 5: Recruiting bees for selected sites (more bees for best s sites), and evaluating the fitness value. In Step 5, the BA conducts search in the neighborhood of the selected sites. More bees will be assigned to search around the best e sites. The bees can either be selected directed based on the value of fitness associated with the sites they are visiting, or the fitness values will be used to determine the probability of the bees being selected. Exploration of the surroundings of the best e sites represents more suitable solutions can be made available through recruiting more bees to follow them than the other selected bees. This differential recruitment mechanism, along with scouting strategy, is the core operation of the BA.
- Step 6: Choosing the fittest bee from each patch. The authors of [42] introduced a constraint at this stage that is for each patch, only the bees with the highest fitness value can be selected to from the next bee population. The purpose of introducing such restriction is to reduce the number of points that are going to be explored.
- Step 7: Assigning the remaining bees to do random search, and evaluating their fitness. In the bee population, the remaining bees are assigned randomly around the search space looking for new possible solution candidates.

- Step 8: Terminating the loop when stopping criterion is met. All aforementioned steps will be executed repeatedly until a stooping criterion is met. At the end of each iteration, the population of bee colony consists of two parts: representatives from each selected patch, and other scout bees performing random searches.

Smart Robot Control using BA. Mobile robots for public service have always been a great research interest both for academic and industry. During the past few years, the study of robot swarm and its related topic has become a hot area. In [43], the authors utilized BA to study a group of reconfigurable mobile robots which are designed to provide daily service in hospital environments for different kinds of tasks such as guidance, cleaning, delivery, and monitoring. The fulfilment of each job requires an associated functional module that can be installed onto various robot platforms via a standard connection interface. Since the classic BA focuses mainly on single-objective functional optimization problems, a variant called binary BA (BBA for short) was proposed by the authors of [43] to deal with the multi-objective multi-constraint combinatorial optimization task. In BBA, a bee is describe as two binary matrixes **MR** and **RH**, standing for how to assign the M tasks to the R robots and the R robots to the H homes, respectively. The size of **MR** is $M \times R$ in which its R columns represent the R robots, while the M missions is represented by the M rows.

The authors evaluated the BBA with an example problem (20 missions, 8 robots, and 4 homes) with a size of $8^{20} \times 4^8 = 2^{76}$ combinations. At first, 12 stochastic solutions are obtained by scout bees through global search in which six elite bees survive after the non-dominated selection. The final experiments demonstrated that BBA is a suitable candidate tool in treating workload balancing issue among a team of swarm robots.

3.3 Glowworm Swarm Optimization (GSO) Algorithm

The glowworm swarm optimization (GSO) algorithm was proposed by the authors of [44]. This algorithm gets its inspiration from the behavior from glowworms and it shares some common points with other population-based algorithms such as ant colony optimization and particle swarm optimization. The agents in GSO are a group of glowworms that carry a luminescent quantity call luciferin. Normally, GSO starts by randomly placing n glowworms in the search space so that they are well distributed. Initially, all glowworms carry an equal quantity of luciferin l_0. Typically, each iteration of GSO algorithm consists of three rules, namely, luciferin-update rule, movement rule, and transition rule. The details are listed as below.

- Luciferin-update phase: It is the process by which the luciferin quantities are modified. The quantities value can either increase, as glowworms deposit luciferin on the current position, or decrease, due to luciferin decay. The luciferin update rule is given as follows [44]:

$$l_i(t+1) = (1 - \rho) \cdot l_i(t) + \gamma \cdot J \cdot [x_i \cdot (t+1)]. \tag{4}$$

where $l_i(t)$ denotes the luciferin level associated with the glowworm i at time t, ρ is the luciferin decay constant $(0 < \rho < 1)$, γ is the luciferin enhancement constant, and $J(x_i(t))$ stands for the value of the objective function at glowworm i's location at time t.

- Movement phase: During this phase, glowworm i chooses the next position j to move to using a bias (i.e., probabilistic decision rule) toward good-quality individual which has higher luciferin value than its own. In addition, based on their relative luciferin levels and availability of local information, the swarm of glowworms can be partitioned into subgroups that converge on multiple optima of a given multimodal function. The probability of moving toward a neighbour is given as follows [44]:

$$p_{ij}(t) = \frac{l_j(t) - l_i(t)}{\sum_{k \in N_i(t)} [l_k(t) - l_i(t)]}. \tag{5}$$

where $j \in N_i(t)$, $N_i(t) = \{j : d_{ij}(t) < r_d^i(t); l_i(t) < l_{js}(t)\}$ is the set of neighbours of glowworm i at time t, $d_{ij}(t)$ denotes the Euclidean distance between glowworms i and j at time t, and $r_d^i(t)$ stands for the variable neighbourhood range associated with glowworm i at time t.

Based on the above equation, the discrete-time model of the glowworm movements can be stated as follows [44]:

$$x_i(t+1) = x_i(t) + s \left[\frac{x_j(t) - x_i(t)}{\|x_j(t) - x_i(t)\|} \right]. \tag{6}$$

where $x_i(t) \in \mathbf{R}^m$ is the location of glowworm i at time t in the m-dimensional real space, $\|\cdot\|$ denotes the Euclidean norm operator, and s (>0) is the step size.

- Neighbourhood range update rule: In addition to the luciferin value update rule that is illustrated in the movement phase, in GSO the glowworms use a radial range (i.e., $(0 < r_d^i \le r_s)$) update rule to explore an adaptive neighbourhood (i.e., to detect the presence of multiple peaks in a multimodal function landscape). Let r_0 be the initial neighbourhood range of each glowworm (i.e., $r_d^i(0) = r_0 \ \forall i$), then the updating rule is given as follows [44]:

$$r_d^i(t+1) = \min\{r_s, \ \max\{0, r_d^i(t) + \beta(n_t - |N_i(t)|)\}\}. \tag{7}$$

where β is a constant parameter, and $n_t \in N$ is a parameter used to control the number of neighbours.

Smart Robot Control using GSO. In [45], the authors built a team of four wheeled-mobile robot named Kinbots, and employ GSO algorithm to fulfill the collective robot control. In the work, they equipped their Kinbots with infrared sensor-based interaction modules which can offer a hardware capability to perform luciferin emission or detection and a behavior of leader-following. In a subsequent study [46], the same authors conducted a preliminary experiment to demonstrate the performance of GSO in robot swarm control. A set of three glowworms (i.e., Kinbots) A, B and C are initially located at the corners of an equilateral triangle (side is 50 cm). Kinbots A and B remained stationary during the study, and emitted a luciferin value of 128 and 60, respectively, while a luciferin value of 40 was glowed by the Kinbot C. At each iteration, the sweep platform performs three tasks: First, homes by turning clock-wise until it make a proper angle with the heading direction of the Kinbots; Second, does a 180° scanning to acquire intensity samples and localize the neighbors; and Third, aligns along the line-of-sight of each neighbor to receive the luciferin value emitted by itself. The sensing phase of the first cycle is completed at 10 s. For simplicity, the authors of [46] introduced a maximum-neighbor selection rule that is a Kinbot chooses to move toward a neighbor which emits maximum luciferin. The simulation results demonstrated the suitability of applying GSO in dealing with the problem of multiple source localization encountered in the domain of robot control.

4 Physics-Based CI

This section introduces a physics-based algorithm, named electromagnetism-like mechanism (EM), which gets its inspiration from electromagnetism theory. The representative application of EM in addressing the issue of smart robot control is also highlighted.

4.1 Electromagnetism-Like Mechanism (EM) Algorithm

Similar to that in the elementary electromagnetism, the authors of [47] regarded teach sample point as a charged particle that is released to a space. In the proposed electromagnetism-like mechanism (EM) approach, the objective function value is associated with the charge of each point which determines the magnitude of attraction/repulsion of the point over the sample population. In other words, the higher the magnitude of attraction, the better the objective function value. Once we get the value of these charges, we can use them to look for a direction, usually obtained through the evaluation of a combination force that exerts on the point via other points, where each point can move toward in the subsequent iterations. Like the electromagnetic forces, by adding vectorially the forces from each of the other points, we can obtain the required force. Typically, the EM algorithm consists of four phases, namely, initialization, local search, total force calculation, and the movement. For the rest of this section, we will explain them in detail.

- Initialization: This procedure is used to sample m points randomly from the feasible domain, which is an n-dimensional hyper-cube. Each coordinate of a point is assumed to be uniformly distributed between the corresponding upper and lower bound. After a point is selected from the space, the objective function value for that point is calculated using the function pointer $f(x)$. The initialization phase ends after identifying m points, and the point that carries the best function value is stored in x^{best}.
- Local search: This phase is used to collect the local information for a point x^i. First, the maximum feasible step length is computed according to the parameter δ. Second, for a given i, the improvement in x^i is searched coordinate by coordinate. For a given coordinate, the point x^i is assigned to a tentative point y for storing the initial information. Third, a random number is chosen as the step length and the point y moves along that direction. If the point y finds a better point within local search iterations, the point x^i is replaced by y and the neighborhood search for points i terminates. Finally, updating the current best point.
- Total force calculation: In EM, according to the superposition principle of electromagnetism theory, point i's power of attraction/repulsion is determined by the charge of each point $i(q^i)$ via Eq. (8) [47].

$$q^i = \exp\left\{-n\frac{f(x^i) - f(x^{best})}{\sum_{k=1}^{m}[f(x^k) - f(x^{best})]}\right\}, \forall i. \tag{8}$$

Unlike the electrical charges, no signs are attached to the charge of an individual point in Eq. (8). Instead, the authors of [47] decided the direction of a particular force between two points based on the comparison of their objective function values. Accordingly, the total force F^i exerted on the point i can be calculated via Eq. (9) [47].

$$F^i = \sum_{j\neq i}^{m} \left\{ \begin{array}{ll} (x^j - x^i)\frac{q^i \times q^j}{\|x^j - x^i\|^2} & \text{if } f(x^j) < f(x^i) \\ (x^i - x^j)\frac{q^i \times q^j}{\|x^j - x^i\|^2} & \text{if } f(x^j) \geq f(x^i) \end{array} \right\}, \forall i. \tag{9}$$

- Movement: After assessing the total force vector F^i, the point i can move in the direction of the force by a random step length as shown in Eq. (10).

$$x^i = x^i + \lambda\frac{F^i}{\|F^i\|}(RNG), \quad \text{for } i = 1, 2, \ldots, m. \tag{10}$$

where λ denotes the random step length and is assumed to be uniformly distributed within $[0, 1]$, and RNG is a vector whose components represent the allowed feasible movement toward the upper bound (u^k) or the lower bound (l^k), respectively, for the corresponding dimension.

Smart Robot Control using EM. Autonomous mobile robots have many possible and promising applications in routine or dangerous tasks such as inter-planetary and underwater exploration, dust suction, and mail delivery. In the area of assistive living robots research, one of the essential problems is the path tracking problem which consists of designing control techniques for the purpose of assuring that the robot can follow a predetermined path. Due to the multivariable and nonlinear nature of this problem, the control strategies design for mobile is a very complex task. In order to deal with this issue, in [48], the authors employed EM to minimize the mobile robot controller's cost function in real time manner. In mobile robot control, the kinematic model is widely used. Normally, in the 2-dimentional Cartesian space, the pose of the robot can be represented through Eq. (11) [48].

$$q = (x, y, \theta)^T. \tag{11}$$

where $(x, y)^T$ is the coordinate of O_r in the reference coordinate system, and θ is the heading direction.

In an ideal condition, the wheels of a mobile robot are assumed non-slidable, and thus, the kinematics model can be expressed through Eqs. (12), (13), and (14) [48].

$$q'(t) = S(q)V(t). \tag{12}$$

$$S(q) = \begin{bmatrix} \cos\theta(t) & -d\sin\theta(t) \\ \sin\theta(t) & d\cos\theta(t) \\ 0 & 1 \end{bmatrix}. \tag{13}$$

$$V(q) = \begin{bmatrix} v(t) \\ w(t) \end{bmatrix}. \tag{14}$$

where $v(t)$ and $w(t)$ denotes the agential and angular velocities of mobile robot, respectively.

The goal of their study is to guarantee that the vehicle follows a pre-determined referent path without the displacement error. Nevertheless, since the destination coordinates vary with time, the authors thus placed a reference virtual robot with the same mathematical shown in Eq. (15) [48] mode on the track.

$$\begin{cases} x'_{ref}(t) = v_{ref}(t)\cos[\theta_{ref}(t)] \\ y'_{ref}(t) = v_{ref}(t)\sin[\theta_{ref}(t)] \\ w'_{ref}(t) = \alpha\theta_{ref}(t)/dt \end{cases} \tag{15}$$

After linearizing Eq. (15) around an operating point and subtracting Eq. (13) from the obtained values, the proposed kinematic model of the error is stated as in Eq. (16) [48].

$$\tilde{x}'(t) = A(t) \cdot \tilde{x}'(t) + B(t) \cdot \widetilde{U}(t). \tag{16}$$

where $\tilde{x} = \hat{x} - \hat{x}_{ref}$ stands for the error to the reference robot, $\widetilde{U}(t) = U(t) - U_{ref}(t)$ represents the control input error, matrices $A(t)$ and $B(t)$ are the Jabocian of (13) in relation to $\tilde{x}(t)$ and $U(t)$, respectively. By utilizing the Euler's method and considering the distribution, a time-variant discrete linear model can be expressed in Eq. (17) [48].

$$\tilde{x}'(k+1) = A(k) \cdot \tilde{x}'(k) + \widetilde{B}(k) \cdot \widetilde{U}(k). \tag{17}$$

where $\widetilde{B}(k)$ is $B(k)$ when adding disturbance $\bar{\tau}_{ed}$ from dynamic system to this system.

Built on aforementioned analysis, the path tracking problem considered in [48] is interpreted in Eq. (18) [48].

$$\bar{M}(q)V'(t) + \bar{V}_m(q, q')V(t) + \bar{F}(V(t)) + \bar{\tau}_d = \bar{B}_\tau. \tag{18}$$

The authors made the comparisons between two algorithm, namely, EM and reference algorithm. From the simulation results, it is observed that the linear and angular velocities of the target mobile robot optimized by EM method have a faster convergence speed. The study of [48] successfully demonstrated that the EM approach present a good performance in minimizing the cost function. The major advantage of employing EM algorithm was, concluded at the end of [48], the ability to provide an effective and simple predictive control strategy.

4.2 Intelligent Water Drops (IWD) Algorithm

The inspiration of intelligent water drops (IWD) algorithm comes from the water drops that flow into rivers, lakes, and seas. The core concept is that gravitational form of the earth drags the water drops in a river to flow towards their final destination. The author of [49] invented this algorithm in 2007. The basic working principles of the IWD algorithm are explained as follows [50].

- In IWD algorithm, each intelligent water drop is assumed to carry a certain amount of the soil, $soil(IWD)$, and to flow at a current velocity, $velocity(IWD)$. Suppose that an IWD wants to move to the node j from its current node i. The velocity of an IWD is updated through Eq. (19) [50] which is based on the amount of the soil on the arc between these two nodes, represented by $soil(i,j)$.

$$vel^{IWD}(t+1) = vel^{IWD}(t) + \begin{cases} \frac{a_v}{b_v + c_v \times soil^{\beta}(i,j)} & \text{if } soil^{\beta}(i,j) \neq \frac{b_v}{c_v} \\ 0 & \text{otherwise} \end{cases} \tag{19}$$

where $vel^{IWD}(t+1)$ stands for the updated velocity of an IWD at the node j, while $a_v, b_v,$ and c_v are constant velocity parameters which are adjustable according to focal problems.

- Meanwhile, the time required for an IWD flowing with the velocity of to move from its current node i to the next j, denoted by $time(i,j; vel^{IWD}(t+1))$, can be calculated via Eq. (20) [50].

$$time\left(i,j; vel^{IWD}\right) = \frac{f(i,j)}{vel^{IWD}}. \tag{20}$$

- In IWD algorithm, a fast water drop removes more soil from the path it flows on than a slower water drop which is similar to what happens in nature: fast rivers always make their beds deeper than the slower ones. Therefore after an IWD travels from node vel^{IWD} to node vel^{IWD}, the soil, vel^{IWD}, on the path between these two nodes is computed through Eq. (21) [50].

$$soil(i,j) = \rho_0 \times soil(i,j) - \rho_n \times \Delta soil(i,j). \tag{21}$$

where ρ_0 and ρ_n are positive numbers fall within the interval of $[0, 1]$. Another important strategy that every IWD must have is how to select its next node. It is assumed in the algorithm, an IWD prefers to take a path which has less amount of soil than the other paths. Let node i be the current position of an IWD, then the probability $p_j^{IWD}(j)$ of choosing the path from node i to node j is calculated using Eq. (22) [50].

$$p_j^{IWD}(j) = \frac{f(soil(i,j))}{\sum_{k \notin vc(IWD)} f(soil(i,k))}. \tag{22}$$

where the inverse of the soil between nodes i and j is computed by through Eq. (23) [50].

$$f(soil(i,j)) = \frac{1}{\varepsilon_s + g(soil(i,j))}. \tag{23}$$

where the parameter of ε_s is a small positive constant number (usually $\varepsilon_s = 0.01$) from preventing a potential division by zero, and $g(soil(i,j))$ is normally used to shift the $soil(i,j)$ on the path connecting nodes i and j towards positive values. The mathematical expression of $g(soil(i,j))$ can be found in Eq. (24) [50].

$$g(soil(i,j)) = \begin{cases} soil(i,j) & \text{if } \min_{l \notin vc(IWD)}(soil(i,l)) \geq 0 \\ soil(i,j) - \min_{l \notin vc(IWD)}(soil(i,l)) & \text{otherwise} \end{cases}.$$

$$(24)$$

- In the proposed algorithm, every created IWD is designed to move from its initial node to the next ones until it finds a solution. At the end of each iteration, the best solution T^{IB} found by the IWDs within such iteration is obtained through Eq. (25) [50].

$$T^{IB} = \arg \max_{\forall T^{IWD}} q(T^{IWD}). \tag{25}$$

- It means that the iteration-best solution T^{IB} is the dominant solution over all other solutions T^{IWD}. At the end of each iteration of the IWD algorithm, the current iteration-best solution T^{IB} is used to update the total best solution T^{IB} as shown in Eq. (26) [50]:

$$T^{TB} = \begin{cases} T^{TB} & \text{if } q(T^{TB}) \geq q(T^{IB}) \\ T^{IB} & \text{otherwise} \end{cases}. \tag{26}$$

Smart Robot Control using IWD. With the recent technological advancement, the development of unmanned vehicular systems, in particular the unmanned combat aerial vehicle (UCAV), have been proved to be beneficial in both military and civilian applications. Nevertheless, the complete benefits of such unmanned systems can only be fulfilled and utilized when their operations could achieve an autonomous level. One of the key requirements for realizing such autonomy is the ability of detecting internal and external changes, and reacting to them in a safe and efficient manner, especially without the intervention from their human operators. Under such circumstance, the trajectory planning becomes a nontrivial task. Typically, the goal of trajectory planning is to generate a space path between an initial location and the desired destination that has an optimal or near-optimal performance under different constraint conditions. In [51], the authors made an attempt to solve this imperative task by utilizing IWD algorithm. According to [51], the focal problem can be briefly described as follows:

Suppose the flight task for an UCAV is from node T^{IB} to node T^{IB}. The mathematical expression of the path connecting these two nodes is shown in Eq. (27) [51].

$$Path = \{o, L_1(x_1, y_{k1}), L_2(x_2, y_{k2}), \ldots, L_{m-1}(x_{m-1}, y_{k(m-1)}), A\}. \tag{27}$$

where $k_i = 1, 2, \ldots, 2n + 1$.

The objective function used in single UCAV trajectory planning can thus be defined according to Eq. (28) [51].

$$J = \int_0^{t_f} \left(\omega_1 \times C^2 + \omega_2 \times h^2 + \omega_3 \times f_{rw} \right) dt. \tag{28}$$

where C denotes the large cross-track deviations from the line joining the starting and target points, f_{rw} penalizes the penetration trajectories that come dangerously close to the known threat sites, and h minimizes the UCAV's altitude above level h. Since only the horizontal path optimization is taken into account in the study of [51], the objective function is thus simplified to the form shown in Eq. (29) [51].

$$J = L_k + \delta \sum_{i=1}^{m-1} \frac{1}{d_{imin}}. \tag{29}$$

where L_k is the flight distance, d_{imin} stands for the distance from the node to the closest threat, δ is an threat avoided coefficient (in general, it is an environmental dependent factor).

The flight distance, from node $a(x_i, y_g)$ in vertical line L_i to node $b(x_{i+1}, y_j)$ in vertical line L_{i+1} can be found in Eq. (30) [51].

$$d_{ab} = \sqrt{ \left(\frac{|AB|}{m} \right)^2 + \left(y_j - y_g \right)^2 }, \quad \text{for } j, g = 1, 2, \ldots, 2n+1. \tag{30}$$

Accordingly, the possible UCAV flight distance can be calculated via Eq. (31) [51]:

$$L_k = \sqrt{ \left(\frac{|AB|}{m} \right)^2 + (y_{ki} - 0)^2 } + \sum_{ki=1}^{m-2} \sqrt{ \left(\frac{|AB|}{m} \right)^2 + \left(y_{k(i+1)} - y_{ki} \right)^2 }$$

$$+ \sqrt{ \left(\frac{|AB|}{m} \right)^2 + \left(y_{k(m-1)} - 0 \right)^2 }. \tag{31}$$

Suppose that the number of threats is q, and each one of them is described by a circle with the center point denoted by (x_i, y_j) and the radius of r_j, the distance between the $node(x_i, y_{ki})$ and the threat j can be computed through Eq. (32) [51].

$$d = \sqrt{ \left(x_i - x_j \right)^2 + \left(y_{ki} - y_j \right)^2 } - r_j. \tag{32}$$

The distance between the $node(x_i, y_{ki})$ and the nearest threat can thus be computed via Eq. (33) [51].

$$d_{imin} = \left[\left\{ \sqrt{(x_i - x_j)^2 + (y_{ki} - y_1)^2} - r_1 \right\}, \ldots, \left\{ \sqrt{(x_i - x_q)^2 + (y_{ki} - y_q)^2} - r_q \right\} \right].$$

$$(33)$$

Since the generated UCAV optimal trajectory using the proposed IWD algorithm is normally difficult to be implemented in real flying environment due to the potential turning points on the optimized trajectory, the authors of [51] further adopted a class of dynamically feasible trajectory smooth strategy named k-trajectory. Finally, a series of case studies were conducted in their study under complicated combating environments. From the experimental results, it can be observed that the proposed IWD algorithm can find a feasible and optimal trajectory for the single UCAV. Meanwhile, the adopted k-trajectory approach is also every effective in smoothing the UCAV trajectory with a small computational load and real-time simulation possibility.

4.3 Gravitational Search Algorithm (GSA)

Gravitational search algorithm (GSA) was originally proposed by Rashedi, Nezamabadi-pour [52]. In GSA, all the individuals can be mimicked as objects with masses. Based on the Newton's law of universal gravitation, the objects attract each other by the gravity force, and the force makes all of them move towards the ones with heavier masses. In addition, each mass of GSA has four characteristics: position, inertial mass, active gravitational mass, and passive gravitational mass. The first one corresponds to a solution of the problem, while the other three are determined by fitness function. The details of GSA can be summarized as follows.

- First, considering a system with N masses (agents) where the ith mass's position is defined by Eq. (34) [52]:

$$X_i = \left(x_i^1, \ldots, x_i^d, \ldots, x_i^n \right), \quad i = 1, 2, \ldots, N. \tag{34}$$

where x_i^d is the position of the ith agent in the dth dimension, and n is the search space's dimension.

- Second, the gravitational force $(F_{ij}^d(t))$ that acting on mass i from mass j at time t can be defined by Eq. (35) [52]:

$$F_{ij}^d(t) = G(t) \frac{M_{pi}(t) \cdot M_{aj}(t)}{R_{ij}(t) + \varepsilon} \left(x_j^d(t) - x_i^d(t) \right). \tag{35}$$

where M_{aj} is the active gravitational mass related to agent j, M_{pi} is the passive gravitational mass related to agent i, $G(t)$ is gravitational constant at time t, ε is a small constant, and $R_{ij}(t)$ is the Euclidian distance between two agents i and j defined by Eq. (36) [52]:

$$R_{ij}(t) = \left\| X_i(t), X_j(t) \right\|_2. \tag{36}$$

- Third, for the purpose of computing acceleration of an agent i, total forces (from a group of heavier masses) can be defined by Eq. (37) [52]:

$$F_i^d(t) = \sum_{j=1, j \neq i}^{N} rand_j F_{ij}^d(t). \tag{37}$$

where $rand_j$ is random number in the interval $[0, 1]$.
- Fourth, based on the total forces, the acceleration of the agent i at time t, and in direction dth, is given by Eq. (38) [52]:

$$a_i^d(t) = \frac{F_i^d(t)}{M_{ii}(t)}. \tag{38}$$

where M_{ii} is the inertial mass of ith agent.
- Fifth, an agent's next velocity can be computed as a fraction of its present velocity added to its acceleration. Both agent i's new velocity and position are given by Eqs. (39) and (40), respectively [52]:

$$v_i^d(t+1) = rand_i \times v_i^d(t) + a_i^d(t). \tag{39}$$

$$x_i^d(t+1) = x_i^d(t) + v_i^d(t+1). \tag{40}$$

where $v_i^d(t)$ and $x_i^d(t)$ are the velocity and the position in dth dimension of agent i at time t, respectively, and $rand_i$ is an uniform random variable in the interval $[0, 1]$ which adds a randomized characteristic to the search.
- Finally, after computing current population's fitness, the gravitational and inertial masses can be updated via Eqs. (41) and (42), respectively [52]:

$$m_i(t) = \frac{fit_i(t) - worst(t)}{best(t) - worst(t)}. \tag{41}$$

$$M_i(t) = \frac{m_i(t)}{\sum_{j=1}^{N} m_j(t)}. \tag{42}$$

where $fit_i(t)$ denotes the fitness value of the agent i at time t, $best(t)$ and $worst(t)$ are defined by Eqs. (43) and (44), respectively [52]:

For a minimization problem:
$$\begin{cases} best(t) = \min_{j \in \{1,\ldots,N\}} fit_j(t) \\ worst(t) = \max_{j \in \{1,\ldots,N\}} fit_j(t) \end{cases}. \tag{43}$$

For a maximization problem:
$$\begin{cases} best(t) = \max_{j \in \{1,\ldots,N\}} fit_j(t) \\ worst(t) = \min_{j \in \{1,\ldots,N\}} fit_j(t) \end{cases}. \tag{44}$$

- Furthermore, to balance the exploration and exploitation of GSA, an agent called k_{best} is employed. It is a function of time which with the initial value k_0 at the beginning and decreasing with time. In GSA, k_0 is normally set to N (i.e., the total number of agents), and k_{best} is decreased linearly. Finally, there will be just one agent applying force to the others. Therefore, the Eq. (37) can be modified as Eq. (45) [52]:

$$F_i^d(t) = \sum_{j \in k_{best}, j \neq i}^{N} rand_j F_{ij}^d(t). \tag{45}$$

where k_{best} is the set of first k agents with the best fitness value and biggest mass.
- Meanwhile, an initial value of G_0 is always allocated to the gravitational constant, G, which will be reduced with time as defined by Eq. (46) [52]:

$$G(t) = G(G_0, t). \tag{46}$$

The steps of implementing GSA can be summarized as follows [52]:

- Step 1: Determining the system environment.
- Step 2: Randomized initialization.
- Step 3: Fitness evaluation of agents.
- Step 4: Updating the parameters, i.e., $G(t), best(t), worst(t)$, and $M_i(t)$ for $i = 1, 2, \ldots, N$.
- Step 5: Calculation of the total force in different directions.
- Step 6: Calculation of acceleration and velocity.
- Step 7: Updating the position of agents.
- Step 8: Repeat Steps 3–7 until the stop criteria is reached. If a specified termination criteria is satisfied stop and return the best solution.

Smart Robot Control using GSA. Although legged robots are slower and more complicated, they have many advantages under certain conditions, such as better adaptability to irregular terrain conditions, and better climbing and obstacle overcrossing capability, which enable them to be more flexible than other types of locomotive mechanisms and can thus be deployed in multitude of dynamically changing situations. While building walking robot, stability is a key factor that needs to be considered since it is fundamental to the overall performance of terrestrial locomotion. In [53], the author made an attempt to use the characteristics of genetic and GSA to generate gait for a hexapod walking robot. The experimental results demonstrated that, in general, the increase of fitness of transformed gaits can be achieved in comparison with the fitness of the initial gait population. At the end of the study, the author of [53] suggested that supplementary mechanisms can be added for compensating the deviation of the robot path from preset trajectory.

5 Experimental Study

5.1 Scenario Settings

Since the main focus of this study falls within AAL robot, we also establish our case study in the context of such theme. It is well known that one of the AAL robots' main applications is in the older person's home, such as healthcare robots [54–58] that are capable of monitoring health (both psychological and medical), detection of falls, medication dispensing, communicating with older people, and assisting physical tasks. Over the decades, the main stream of the existing studies are focused on homogeneous robots (i.e., all individual behaviours are the same), since it becomes easier for one robot to model other robots in the group, which in turn can enhance the whole group's robustness against the failure of individual robot [59]. Nevertheless, there is much to concern the heterogeneous robot groups [60–65] in order to provide more complicated services, such as task assignment, distribution, and rescue. Other examples like robot soccer [66, 67] which involves a variety of agent types, would more easily be solved with a suitable heterogeneous oriented control approach. As a result, this study focuses on a heterogeneous robot group in the AAL environment (i.e., smart home) with each responsible for separate tasks, e.g. cooking, cleaning, maintenance, entertainment, etc. Under this circumstance, when an elderly person suddenly falls down, different functionality equipped robots will form a swarm and try to help by pulling him/her to a comfortable location.

5.2 Core Challenges Faced by the Experimental Study

In general, a heterogeneous robot group consists of robots with different designs or functionalities that usually complement each other in order to complete tasks efficiently [68]. Unfortunately, turning such concept (i.e., a heterogeneous swarm of robots) into reality is not very easy. The core challenge for that is how the underlying control schemes will operate, which must deal with the level of behavioural control and the hardware design [68].

Normally, the main inspiration for such swarm robots' behavioural control comes from the observation of social animals (e.g., ants, birds, herd, and fish), since they are capable of self-organizing, decentralization, and emergence of collective behaviour [69]. For example, the behaviour of ants [70] has inspired a number of researchers in collective robotics, such as [71–75]. An interesting initiative is "Swarmanoid" project [68, 76], which provides a modular framework for heterogeneous robot groups' control. Also, a number of researchers (e.g., [65]) have explored heterogeneity at the hardware level (e.g., different sets of sensors or effectors). As such, in this work, we make an attempt to study the effects of a newly developed social animal inspired algorithm (i.e., grey wolf optimizer) in controlling a heterogeneous swarm of AAL robots.

5.3 Employed Novel CI Algorithm

In this case study, we will employ a newly developed algorithm called grey wolf optimizer (GWO) [77] which is inspired by the grey wolves hunting and searching behaviours. In [77], the authors classified the wolves into 4 groups, i.e., alphas, betas, deltas, and omegas. In addition, they assumed that among the groups, alpha (the fittest solution), beta, and delta have better knowledge about the potential location of prey [77]. Overall, the GWO can be seen as a two-stage method, i.e., encircling the prey during the hunting process using hyper-cubes framework and then employing an intensive local search mechanism for optimization. Like many other novel CI algorithms, GWO also includes a balance between exploitation/exploration. This offers the advantage of enhanced search ability while maintaining adequate exploitation capability.

In the following, we describe the steps to be taken for obtaining an efficient implementation of GWO.

- Step 1: Generate the initial the grey wolf population, $X_i(i = 1, 2, \ldots, n)$.
- Step 2: Initialize the algorithm parameters $(a, A, \text{ and } C)$ via Eq. (47) [77]:

$$
\begin{aligned}
\overrightarrow{A} &= 2\overrightarrow{a} \cdot \overrightarrow{r}_1 - \overrightarrow{a} \\
\overrightarrow{C} &= 2 \cdot \overrightarrow{r}_2.
\end{aligned}
\tag{47}
$$

where $\overrightarrow{A}$ and $\overrightarrow{C}$ are coefficient vectors, the components of $\overrightarrow{a}$ are linearly decreased from 2 to 0 over the course of iterations, and $\overrightarrow{r}_1$ and $\overrightarrow{r}_2$ are random vectors in $[0, 1]$.
- Step 3: Evaluating the fitness value, i.e., $X_\alpha, X_\beta,$ and X_δ.
- Step 4: Position correction-cooperation between current search agents by Eqs. 48, and 49, respectively [77]:

$$
\begin{cases}
\vec{D}_\alpha = \left| \vec{C}_1 \cdot \vec{X}_\alpha - \vec{X} \right| \\
\vec{D}_\beta = \left| \vec{C}_2 \cdot \vec{X}_\beta - \vec{X} \right| . \\
\vec{D}_\delta = \left| \vec{C}_3 \cdot \vec{X}_\delta - \vec{X} \right|
\end{cases}
\tag{48}
$$

$$
\begin{cases}
\vec{X}_1 = \vec{X}_\alpha - \vec{A}_1 \cdot \left(\vec{D}_\alpha \right) \\
\vec{X}_2 = \vec{X}_\beta - \vec{A}_2 \cdot \left(\vec{D}_\beta \right) . \\
\vec{X}_3 = \vec{X}_\delta - \vec{A}_3 \cdot \left(\vec{D}_\delta \right)
\end{cases}
\tag{49}
$$

where $\vec{D}$ is the distance of each candidate solution from the prey, $\vec{X}_\alpha$, $\vec{X}_\beta$, and $\vec{X}_\delta$ are the positions vector of the prey, t represents the current iteration, and $\vec{X}$ indicates the position vector of a grey wolf.

- Step 5: Updating the best location of the hunting wolves through Eq. (50) [77]:

$$
\vec{X}(t+1) = \frac{\vec{X}_1 + \vec{X}_2 + \vec{X}_3}{3}.
\tag{50}
$$

- Step 6: Evaluating the stopping criteria. If yes, generate output; otherwise, go back to Step 2.

Although the GWO is designed in a very simple manner, i.e., only three main parameters need to be adjusted, each parameter has its own functionalities. For example, the objective for parameters a and A is to find a reasonable balance between a too narrow focus of the search process, which in the worst case may lead to stagnation, and a too weak guidance of the search, which can cause excessive exploration, i.e., when $|A| < 1$, the wolves will attack towards the prey; otherwise, the wolves keep searching for prey. In addition, parameter C has two features: First, it provides random weights for prey in order to stochastically emphasize $(C > 1)$ or deemphasize $(C < 1)$ the effect of prey in defining the distance; Second, it represents the effect of obstacles to approaching prey in nature [77].

5.4 Testing Results

To solve the focal heterogeneous robot swarm control problem, GWO is utilized to optimize the group performance. We have four kinds of robots, namely, inchworm-type, wheel type, crocodile-type, and caterpillar-type (see Fig. 1 for illustration).

Initially only two types robots were employed and the time of completing the pulling task is recorded. Later on, another type of robot was added and the duration of task finishing is again recorded. Finally, we add one more type of robot into the

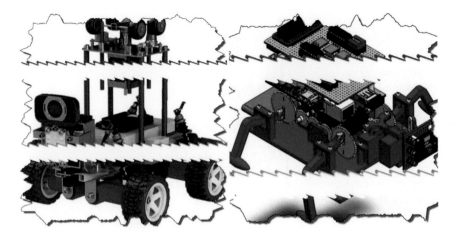

Fig. 1 The schema of wheel-type and crocodile-type robots

No. of Robots	Type of Robots				Time Required (in second)	Distance Traveled (in meter)	Task Accuracy (%)
	Inchworm-type	Wheel-type	Crocodile-type	Caterpillar-type			
2	✔	✔			1200	27	37
2	✔		✔		1600	35	28
2	✔			✔	1100	30	39
⋮	⋮	⋮	⋮	⋮	⋮	⋮	⋮
3	✔	✔	✔		840	45	45
⋮	⋮	⋮	⋮	⋮	⋮	⋮	⋮
4	✔	✔	✔	✔	600	60	75

Legend: ✔ denotes the selected robot.

Fig. 2 Preliminary experimental results

recue group and write down the cost of time in this situation. The experimental results are show in Fig. 2. As we can see, different time consumption and task accuracy can be achieved in distinct robot deployment. In general, according to our experiment, the more robots involved, the less time required and the higher task accuracy obtained. Nevertheless, we believe this is scenario dependent and interest readers need to tailor their own arrangement to reach the best overall system performance.

6 Future Work

The limitations of the present work are twofold: On one hand, a widespread survey of the applications of innovative CI presents a difficult task, because of the extensive background knowledge that is required during the process of collection, study, and classification of these publications. Although acknowledging a limited background knowledge, this research makes a brief overview of literature concerned with using biology- and physics-based innovative CI algorithms to deal with robot control. On the other hand, there are also some algorithms that falls under the other categories such as chemistry-, mathematics-based innovative CI. However, the present study does not take them into account. This would be an immediately research direction that need to be considered in future study. Additionally, the presented experimental study is limited to one specific AAL scenario. The future work of this study can be expanded to more general situations with the assistance more comparable novel CI algorithms.

7 Conclusion

Assistive living robot development has for sure not reached its limits and there is still a lot of work need to be done to bridge the gap between academic research and real-world deployment. In this work, we make an attempt to take a quick investigation of several innovative CI techniques and their applications in the area of robot control which is, from the author's perspective, the key to a successfully AAL robot implementation. As the main contribution of this work, we aim at providing a picture about what emerging CI algorithms are being developed and applied to smart robot control area. It is believed that this overview can, through the scattered literature, offer a useful research guide to other scholars who share the similar research interests. The presented demo case also illustrates how these novel CI algorithms can be applied to real world AAL cases.

References

1. Dario, P., et al.: Robotics for medical application. In: Robotics & Automation Magazine (1996), pp. 44–56
2. Dario, P., et al.: MOVAID: a personal robot in everyday life of disabled and elderly people. Technol. Disabil. **10**, 77–93 (1999)
3. Dario, P., Guglielmelli, E., Laschi, C.: Humanoids and personal robots: design and experiments. J. Rob. Syst. **18**(12), 673–690 (2001)
4. JinYoo, S.: Adaptive neural tracking and obstacle avoidance of uncertain mobile robots with unknown skidding and slipping. Inf. Sci. **238**, 176–189 (2013)
5. Jaradat, M.A.K., Al-Rousan, M., Quadan, L.: Reinforcement based mobile robot navigation in dynamic environment. Rob. Comput. Integr. Manuf. **27**, 135–149 (2011)

6. Navarro-Guerrero, N., et al.: Real-world reinforcement learning for autonomous humanoid robot docking. Rob. Auton. Syst. **60**, 1400–1407 (2012)
7. Maravall, D., Lope, J.D., Domínguez, R.: Coordination of communication in robot teams by reinforcement learning. Rob. Auton. Syst. **61**, 661–666 (2013)
8. Herrero-Pérez, D., et al.: Fuzzy uncertainty modeling for grid based localization of mobile robots. Int. J. Approximate Reasoning **51**, 912–932 (2010)
9. Miljkovic, Z., et al.: Neural network reinforcement learning for visual control of robot manipulators. Expert Syst. Appl. **40**, 1721–1736 (2013)
10. Jolly, K.G., Kumar, R.S., Vijayakumar, R.: Intelligent task planning and action selection of a mobile robot in a multi-agent system through a fuzzy neural network approach. Eng. Appl. Artif. Intell. **23**, 923–933 (2010)
11. Motlagh, O., et al.: An expert fuzzy cognitive map for reactive navigation of mobile robots. Fuzzy Sets Syst. **201**, 105–121 (2012)
12. Al-Dabbagh, R.D., et al.: System identification and control of robot manipulator based on fuzzy adaptive differential evolution algorithm. Adv. Eng. Softw. **78**, 60–66 (2014)
13. Kondo, T., Ito, K.: A reinforcement learning with evolutionary state recruitment strategy for autonomous mobile robots control. Rob. Auton. Syst. **46**, 111–124 (2004)
14. Son, J.-H., Choi, Y.-C., Ahn, H.-S.: Bio-insect and artificial robot interaction using cooperative reinforcement learning. Appl. Soft Comput. **18**(6), 1127–1141 (2014)
15. Wu, Y.-H., et al.: Designing an assistive robot for older adults: the ROBADOM project. IRBM **34**, 119–123 (2013)
16. Rashidi, P., Mihailidis, A.: A survey on ambient-assisted living tools for older adults. IEEE J. Biomed. Health Inform. **17**(3), 579–590 (2013)
17. Feil-Seifer, D., Matarić, M.J.: Defining socially assistive robotics. In: Proceedings of the 2005 IEEE 9th International Conference on Rehabilitation Robotics, 28 June–1 July 2005, Chicago, IL, USA
18. Graf, B.: Care-O-bot (2014). http://www.care-o-bot.de/en/care-o-bot-3.html. Accessed on 30 July 2015
19. Graf, B., Hans, M., Schraft, R.D.: Care-O-bot II: development of a next generation robotic home assistant. Auton. Robots **16**, 193–205 (2004)
20. Graf, B., Parlitz, C., Hägele, M.: Robotic home assistant Care-O-bot 3 product vision and innovation platform. In: Jacko, J.A. (ed.) Human-Computer Interaction, Part II, (HCII 2009), LNCS 5611. Springer, Berlin (2009), pp. 312–320
21. RIKEN-TRI Collaboration Center: RIBA (2015) http://rtc.nagoya.riken.jp/RIBA/index-e.html. Accessed on 30 July 2015
22. Mukai, T., et al.: Development of a nursing-care assistant robot RIBA that can lift a human in Its arms. In: The 2010 IEEE/RSJ International Conference on Intelligent Robots and Systems, 18–22 Oct 2010, Taipei, Taiwan (2010)
23. Kuindersma, S.R., et al.: Dexterous mobility with the uBot-5 mobile manipulator. In: International Conference on Advanced Robotics (ICAR), June 2009 (2009), pp. 1–7
24. Xu, J., et al.: Enhanced bimanual manipulation assistance with the personal mobility and manipulation appliance (PerMMA). In: The 2010 IEEE/RSJ International Conference on Intelligent Robots and Systems, 18–22 Oct 2010, Taipei, Taiwan (2010)
25. Wang, H., et al.: The personal mobility and manipulation appliance (PerMMA): a robotic wheelchair with advanced mobility and manipulation. In: The 34th Annual International Conference of the IEEE EMBS, San Diego, California USA, 28 Aug–1 Sept 2012 (2012)
26. Cooper, R.A., et al.: Personal mobility and manipulation appliance-design, development, and initial testing. Proc. IEEE **100**(8), 2505–2511 (2012)
27. Sato, M., Sugiyama, A., Ohnaka, S.I.: Auditory system in a personal robot, PaPeRo. In: 2006 Digest of technical Papers International Conference on Consumer Electronics (ICCE 06), 7–11 Jan 2006 (2006), pp. 19–20
28. Sato, M., et al.: A single-chip speech dialogue module and its evaluation on a personal robot. PaPeRo-mini in IEEE International Conference on Acoustics, Speech and Signal Processing (ICASSP), 19–24 April 2009, Taipei, Taiwan (2009), pp. 3697–3700

29. Fujiwara, N., Hagiwara, Y., Choi, Y.: Development of a learning support system with PaPeRo. In: The 12th International Conference on Control, Automation and Systems, 17–21 Oct 2012, Jeju Island, Korea (2012), pp. 1912–1915

30. Hosoda, Y., et al.: Collision-avoidance algorithm for human-symbiotic robot. In: International Conference on Control, Automation and Systems 2010, 27–30 Oct 2010, Gyeonggi-do, Korea (2010), pp. 557–561

31. Hosoda, Y., et al.: Basic design of human-symbiotic robot EMIEW, in Proceedings of the 2006 IEEE/RSJ International Conference on Intelligent Robots and Systems, 9–15 Oct 2006, Beijing, China (2006), pp. 5079–5084

32. HITACHI. Robotics: EMIEW 2 (2014). http://www.hitachi.com/rd/portal/research/robotics/emiew2_01.html. Accessed on 30 Jul 2015

33. Falconer, J.: HOSPI-R drug delivery robot frees nurses to do more important work (2013). http://www.gizmag.com/panasonic-hospi-r-delivery-robot/29565/. Accessed on 30 July 2015

34. Murai, R., et al.: A novel visible light communication system for enhanced control of autonomous delivery robots in a hospital. In: IEEE/SICE International Symposium on System Integration (SII), 16–18 Dec 2012, Kyushu University, Fukuoka, Japan (2012), pp. 510–516

35. Marwala, T., Lagazio, M.: Militarized conflict modeling using computational intelligence. Springer, London, UK (2011). ISBN 978-0-85729-789-1

36. Xing, B., Gao, W.-J.: Innovative computational intelligence: a rough guide to 134 clever algorithms. Springer International Publishing Switzerland, Cham (2014). ISBN 978-3-319-03403-4

37. Passino, K.M.: Biomimicry of bacterial foraging for distributed optimization and control. IEEE Control Syst. Manage. 22(3), 52–67 (2002)

38. Boussaïd, I., Lepagnot, J., Siarry, P.: A survey on optimization metaheuristics. Inf. Sci. 237, 82–117 (2013)

39. El-Abd, M.: Performance assessment of foraging algorithms vs. evolutionary algorithms. Inf. Sci. 182, 243–263 (2012)

40. Tang, W.J., Wu, Q.H.: Biologically inspired optimization: a review. Trans. Inst. Meas. Contr. 31(6), 495–515 (2009)

41. Hossain, M.A., Ferdous, I.: Autonomous robot path planning in dynamic environment using a new optimization technique inspired by bacterial foraging technique. Rob. Auton. Syst. 64, 137–141 (2015)

42. Pham, D.T., et al.: The bees algorithm—A novel tool for complex optimisation problems. In: Second International Virtual Conference on Intelligent Production Machines and Systems (IPROMS). Elsevier, Oxford (2006), pp. 454–459

43. Xu, S., et al.: Bio-inspired binary bees algorithm for a two-level distribution optimisation problem. J. Bionic Eng. 7, 161–167 (2010)

44. Krishnanand, K.N., Ghose, D.: Glowworm swarm optimization for simultaneous capture of multiple local optima of multimodal functions. Swarm Intell. 3, 87–124 (2009)

45. Krishnanand, K.N., Ghose, D.: Detection of multiple source locations using a glowworm metaphor with applications to collective robotics. In: IEEE Swarm Intelligence Symposium (SIS). IEEE (2005), pp. 84–91

46. Krishnanand, K.N., et al.: Glowworm-inspired robot swarm for simultaneous taxis towards multiple radiation sources. In: IEEE International Conference on Robotics and Automation (ICRA), May, Orlando, Florida, USA. IEEE (2006), pp. 958–963

47. Birbil, Şİ., Fang, S.-C.: An electromagnetism-like mechanism for global optimization. J. Global Optim. 25, 263–282 (2003)

48. Wang, Y., et al.: A model predictive control strategy for path-tracking of autonomous mobile robot using electromagnetism-like mechanism. In: International Conference on Electrical and Control Engineering (ICECE). IEEE (2010), pp. 96–100

49. Shah-Hosseini, H.: Problem solving by intelligent water drops. In: IEEE Congress on Evolutionary Computation (CEC), 25–28 Sept 2007. IEEE (2007), pp. 3226–3231

50. Shah-Hosseini, H.: Intelligent water drops algorithm: a new optimization method for solving the multiple knapsack problem. Int. J. Intell. Comput. Cybern. 1(2), 193–212 (2008)

51. Duan, H., Liu, S., Wu, J.: Novel intelligent water drops optimization approach to single UCAV smooth trajectory planning. Aerosp. Sci. Technol. **13**, 442–449 (2009)
52. Rashedi, E., Nezamabadi-pour, H., Saryazdi, S.: GSA: a gravitational search algorithm. Inf. Sci. **179**, 2232–2248 (2009)
53. Seljanko, F.: Hexapod walking robot gait generation using genetic-gravitational hybrid algorithm. In: 15th International Conference on Advanced Robotics, 20–23 June 2011, Tallinn University of Technology, Tallinn, Estonia. IEEE (2011), pp. 253–258
54. Robinson, H., MacDonald, B., Broadbent, E.: The role of healthcare robots for older people at home: a review. Int. J. Social Robot. **6**(4), 575–591 (2014)
55. Broadbent, E., et al.: Attitudes towards health-care robots in a retirement village. Australas. J. Ageing **31**(2), 115–120 (2012)
56. Andrade, A.O., et al.: Bridging the gap between robotic technology and health care. Biomed. Signal Process. Control **10**, 65–78 (2014)
57. Harrefors, C., Axelsson, K., Sävenstedt, S.: Using assistive technology services at differing levels of care: healthy older couples' perceptions. J. Adv. Nurs. **66**(7), 1523–1532 (2010)
58. Broadbent, E., Stafford, R., MacDonald, B.: Acceptance of healthcare robots for the older population: review and future directions. Int. J. Social Robot. **1**, 319–330 (2009)
59. Potter, M.A., Meeden, L.A., Schultz, A.C.: Heterogeneity in the coevolved behaviors of mobile robots: the emergence of specialists. In: Proceedings of the 7th International Conference on Artificial Intelligence (2011), pp. 1337–1343
60. Parker, L.E.: ALLIANCE: an architecture for fault tolerant, cooperative control of heterogeneous mobile robots. In: Proceedings of the IEEE/RSJ International Conference on Intelligent Robots and Systems (1994), pp. 776–783
61. Haque, M.A.: Biologically inspired heterogeneous multi-agent systems. In: School of Electrical and Computer Engineering, Georgia Institute of Technology (2010)
62. Wurm, K.M., et al.: Coordinating heterogeneous teams of robots using temporal symbolic planning. Auton. Robots **34**, 277–294 (2013)
63. Baca, J., Ferre, M., Aracil, R.: A heterogeneous modular robotic design for fast response to a diversity of tasks. Rob. Auton. Syst. **60**, 522–531 (2012)
64. Lope, J.D., Maravall, D., Quiñonez, Y.: Response threshold models and stochastic learning automata for self-coordination of heterogeneous multi-task distribution in multi-robot systems. Robot. Auton. Syst. **61**, 714–720 (2013)
65. Brunete, A., et al.: A behaviour-based control architecture for heterogeneous modular, multi-configurable, chained micro-robots. Rob. Auton. Syst. **60**, 1607–1624 (2012)
66. Stone, P.: Intelligent autonomous robotics: a robot soccer case study. Morgan & Claypool, (2007). www.morganclaypool.com. ISBN 1598291262
67. Lauer, M., et al.: Cognitive concepts in autonomous soccer playing robots. Cogn. Syst. Res. **11**, 287–309 (2010)
68. Dorigo, M., et al.: Swarmanoid: a novel concept for the study of heterogeneous robotic swarms. In: IEEE Robotics & Automation (2013), pp. 60–71
69. Dorigo, M., Stützle, T.: Ant Colony Optimization. The MIT Press, Cambridge (2004). ISBN 0-262-04219-3
70. Deneubourg, J.L., et al.: The dynamics of collective sorting robot-like ants and ant-like robots. In: Proceedings of 1st Conference on Simulation of Adaptive Behavior (1991)
71. Kube, C.R., Bonabeau, E.: Cooperative transport by ants and robots. Rob. Auton. Syst. **30**, 85–101 (2000)
72. Holland, O., Melhuish, C.: Stigmergy, self-organization, and sorting in collective robotics. Artif. Life **5**, 173–202 (1999)
73. Caro, G.D.: A Society of Ant-like Agents for Adaptive Routing in Networks. Universite Libre de Bruxelles, Brussels, Belgium (2002)

74. Dorigo, M.: Swarms of self-assembling robots. In: Weyns, D., Brueckner, S.A., Demazeau, Y. (eds) EEMMAS 2007, LNAI 5049. Springer, Berlin (2008), pp. 1–2
75. Dorigo, M., et al.: SWARM-BOT: design and implementation of colonies of self-assembling robots. In: Yen, G.Y., Fogel, D.B. (eds) Computational Intelligence: Principles and Practice. IEEE Computational Intelligence Society, New York (2006), pp. 103–135
76. Ferrante, E.: A Control Architecture for a Heterogeneous Swarm of Robots: The Design of a Modular Behavior-Based Architecture. Universite Libre de Bruxelles (2009)
77. Mirjalili, S., Mirjalili, S.M., Lewis, A.: Grey wolf optimizer. Adv. Eng. Softw. **69**, 46–61 (2014)

Author Biography

Bo Xing, DIng, is an Associate Professor at the Department of Computer Science, School of Mathematical and Computer Science, Faculty of Science and Agriculture, University of Limpopo, South Africa. He was a senior lecturer under the division of Center for Asset Integrity Management (C-AIM) at the Department of Mechanical and Aeronautic Engineering, Faculty of Engineering, Built Environment and Information Technology, University of Pretoria, South Africa. Dr. Xing earned his DIng degree (Doctorate in Engineering with a focus on soft computing and remanufacturing) in the early 2013 from the University of Johannesburg, South Africa. He also obtained his B.Sc. and M.Sc. degree both in Mechanical Engineering from the Tianjin University of Science and Technology, P.R. China, and the University of KwaZulu-Natal, South Africa, respectively.

He was a scientific researcher at the Council for Scientific and Industrial Research (CSIR), South Africa. He has published more than 50 research papers in books, international journals, and international conference proceedings. His current research interests lie in applying various nature-inspired computational intelligence methodologies towards big data analysis, miniature robot design and analysis, advanced mechatronics system, and e-maintenance.

Valorization of Assistive Technologies for Cognition: Lessons and Practices

Stéphanie Pinard, Kevin Bouchard, Yannick Adelise,
Véronique Fortin, Hélène Pigot, Nathalie Bier and Sylvain Giroux

1 Introduction

Cognitive rehabilitation is an important part of the services delivered by the healthcare system and requires increasing human and financial resources from the organizations concerned [1]. It can be defined as systematic therapeutic activities aimed at helping persons with cognitive impairment to regain their functional autonomy. The consequences of cognitive impairment are numerous and include difficulty remembering and learning new things, concentrating, and making decisions that affect one's everyday life. A cognitive impairment is generally classified by order of magnitude ranging from mild or moderate to severe. It can be caused by various types of disease or trauma such as stroke, brain injury, dementia (e.g.

S. Pinard (✉) · K. Bouchard · Y. Adelise · V. Fortin · H. Pigot · S. Giroux
DOMUS, Université de Sherbrooke, Sherbrooke, Québec J1H 1H9, Canada
e-mail: Stephanie.Pinard@usherbrooke.ca

K. Bouchard
e-mail: Kevin.Bouchard@usherbrooke.ca

Y. Adelise
e-mail: Yannick.Adelise@usherbrooke.ca

V. Fortin
e-mail: Veronique.Fortin2@usherbrooke.ca

H. Pigot
e-mail: Helene.Pigot@usherbrooke.ca

S. Giroux
e-mail: Sylvain.Giroux@usherbrooke.ca

N. Bier
Centre de recherche, Institut de gériatrie de Montréal,
Université de Montréal, Montréal, Québec H3C 3J7, Canada
e-mail: Nathalie.Bier@umontreal.ca

© Springer International Publishing Switzerland 2016
K.K. Ravulakollu et al. (eds.), *Trends in Ambient Intelligent Systems*,
Studies in Computational Intelligence 633, DOI 10.1007/978-3-319-30184-6_3

Alzheimer's disease), multiple sclerosis, tumor, mental illness, etc. Cognitive rehabilitation can take many forms including compensating for the cognitive deficits with external aids (e.g. using a calendar or smart phone to manage a schedule [2]), modifying environmental factors (e.g. turning the television off when doing a complex task such as cooking) and even training the cognitive deficits specifically (e.g. training attentional capacities). One of the limitations of traditional cognitive rehabilitation is its cost. Many authors have noted that the global cost of care for those with cognitive impairment is becoming unsustainable [3, 4]. Furthermore, many of the diseases or traumas that cause cognitive impairment are projected to increase through the next few decades. For example, it is estimated that Canadians living with Alzheimer's disease will grow from 747,000 in 2011 to 1.4 million by 2031 [5]. Similarly, it is estimated that 10 million people will be affected annually by Traumatic Brain Injury (TBI) and, by the year 2020, it will surpass many diseases as the major cause of death and disability [6]. The consequences of these projections are not only monetary. To improve the lives of persons with cognitive impairments, rehabilitation and support services require a lot of qualified human resources. However, in the current healthcare context, such human resources are difficult to provide [7].

Many researchers are now trying to develop Assistive Technologies for Cognition (ATC) with the goal of improving the quality of life of natural caregivers and addressing the resource issues [8]. This objective is backed by the American Congress of Rehabilitation Medicine [9], which encourages the use of external compensatory aids and maintains that it should be standard practice in cognitive rehabilitation. Interventions with ATC offer entirely new treatment methods that can reinforce a person's residual intrinsic abilities. These interventions provide alternative means by which activities can be completed and extrinsically supported. Without this support, those functional activities might not be performed [10]. In addition, ATCs provide continuous support over time that cannot be supplied by healthcare providers, families and close friends. ATCs also support repetitive and continuous interventions without any additional human resource costs, thus increasing the intensity of the rehabilitation. This intensity, which might be obtained in different settings (home, grocery store, etc.), is important to ensure the success of the intervention with persons suffering from cognitive impairments [11].

Work has been done on many projects to enable such assistive services, particularly with regard to the exploitation of smart homes [12–15]. A smart home is a standard house in which different kinds of sensors and effectors are introduced, generally in a non-invasive fashion [16], in order to provide continuous assistance throughout all Activities of Daily Living (ADLs) performed by the residents. Smart homes are generally designed to be a generic assistive service that could ensure safety, autonomy and well-being at all times. However, there are still many challenges to meet for smart homes to become a realistic solution in rehabilitation [17]. Moreover, cognitive rehabilitation generally differs greatly from one person to the next and thus requires adaptable solutions. That is why, recently, many researchers have turned their attention to simpler, yet more specific, assistive technologies [2]. For example, some rely on a single device such as a pager [18] or personal digital assistant [19]. Another example, developed at our laboratory, DOMUS (Laboratoire de DOMotique de l'Université de Sherbrooke: www.domus.usherbrooke.ca), is called AP@LZ [20]. AP@LZ is a personalized electronic organizer which was conceived to reduce the impact of memory losses in persons with Alzheimer's disease and improve their autonomy. A growing body of literature suggests that ATCs are efficient and effective for improving independence and life participation for people with cognitive impairments [2, 10, 21, 22]. However, ATCs implementation in the field of cognitive rehabilitation remains a real challenge [23]. These intervention tools are still not common in clinics and only a few assistive technologies developed in research reach the commercialization phase [24].

This chapter discusses difficulties with regard to the valorization of ATCs and their widespread adoption. It is divided into three parts. First, it looks at why the implementation of cognitive assistants in cognitive rehabilitation remains difficult. This first part describes the issues at the level of project management, planning, experiments and multidisciplinarity. Second, it presents a specific case of ATC developed at the DOMUS laboratory and analyzes its successes and failures. This second section outlines the various lessons learned through more than 10 years of developing ATCs. Finally, the last section provides a reflective tool to optimize the implementation of cognitive assistants based on the literature. This last section is intended as a general discussion with a view to improving the valorization of ATCs.

2 Challenges Regarding the Valorization of ATCs

Before addressing the challenges related to the valorization of ATCs, it is important to present a definition of what they are. The term ATCs is only broadly defined in the literature. According to Scherer et al. [25], ATCs can refer to very familiar, simple, low-cost devices used by people with and without disabilities to support memory, organization or other cognitive functions, such as paper planners, calendars, alarm clocks, wristwatches and shopping lists. However, in this chapter, we use this term to refer to highly technical devices that compensate for cognitive impairments across environments and task domains. The valorization of ATCs has proven very difficult for researchers around the world and very few concrete products exist despite the amount of literature about ATCs. Valorization can be defined as the transformation of knowledge into concrete new products, services, or processes. The valorization of ATCs thus implies the exploitation of computer science knowledge and models in the development of products in concordance with the knowledge of rehabilitation and professionals. In our case, by valorization we also mean the transition from prototype to adopted product. This step seems to be the most difficult to achieve.

Experts in the field of cognitive rehabilitation have strongly suggested the use of ATCs [2, 10, 21–23, 26]. However, even if the majority of potential users show an interest in assistive devices [23], many challenges preclude their widespread valorization [23]. For the purpose of this chapter, we have divided those challenges into three main categories that will be described in detail in the present section, i.e. issues related to (1) the nature of innovation, (2) the stakeholders, and (3) the commercialization and marketing of ATCs.

2.1 Issues Related to the Nature of Innovation

The first challenge regarding the implementation of ATCs is related to the nature of innovation itself. According to Mulgan et al. [27], innovation is defined as "New Ideas that Work". An invention that gets adopted and used by a community is what defines an innovation. Thus the concept of innovation goes beyond the ideas and products of research in order to get to the step of their implementation [28, 29]. Moreover, innovation involves the introduction of new practices [30, 31] since practical use is necessary for an invention to become an innovation [32]. The term implementation is therefore defined as a description of "how" the transition from invention to innovation will occur. The very nature of the innovations in the quest to develop ATCs is hybrid, that is, the innovation is simultaneously technological [33] and social [34]. Technological innovations are defined as the introduction of new products or existing products with significantly improved technological characteristics [33]. They are often also regarded as something which was not obvious, as per the rule-of-thumb in the patenting system.

Social innovations can, in turn, be defined as new treatment settings that aim to improve the living conditions of patients [31, 34]. To implement ATCs in clinical settings, clinicians must adjust or even transform their practice. Once developed, ATCs need to go through a thorough process before being provided to users. They can hardly be sold or prescribed to clients with cognitive deficits without a complete assessment of clients' needs by a qualified professional and without several teaching sessions delivered by the same professional. This is in part due to the characteristics of these clients. More specifically, the complexity of matching a person with a technology does not stem only from the individual's unique combination of physical, sensory, and cognitive abilities. In addition, people's expectations of and reactions to technologies are complex and highly individualized. Predispositions to technology usage also depend on one's temperament/personality, subjective quality of life/well-being, views of physical capabilities, expectations for future functioning, and financial/social/environmental support for technology use [25]. Thus, to encourage the use of ATCs, rehabilitation services and technological services (maintenance, updates, etc.) are required and many clinical settings do not possess the necessary resources to adopt new practices related to technology. Today, there are various frameworks supporting the changes in clinical practice (such as the NICE Model [35]) and others supporting the commercialization of ATCs, but none of the frameworks in the literature takes into account the 'hybrid' nature of this type of innovation.

2.2   Issues Related to the Stakeholders

The second very important consideration regarding the valorization of ATCs is the consequence of the multidisciplinary nature of these projects. A variety of groups are concerned by the arrival of assistive technologies in cognitive rehabilitation, including healthcare professionals, healthcare administrators, patients and their caregivers. Moreover, computer scientists and engineers generally have a very different vision of how such ATCs should be conceived compared to the vision of healthcare personnel. This is often due to a lack of understanding of the parties involved. Thus, to promote the valorization of ATCs, it is essential to take into account the specific challenges related to the different stakeholders.

2.2.1   Healthcare Professionals, Administrators and the Organizational System Aspect

Many professionals are involved in the day-to-day care of persons with cognitive deficits, including occupational therapists, physical therapists, nurses, neuropsychologists and physicians. This population is currently growing [10], which means a demand for more time and work from the healthcare professionals. de Joode et al. [23] conducted a study which aimed to provide recommendations for the successful implementation of Assistive Technology (AT) in cognitive rehabilitation by investigating the attitudes towards AT of professionals, individuals with acquired brain injury (ABI) and their caregivers. According to that exploratory study, the majority of the professionals working in cognitive rehabilitation showed an interest in AT and were willing to use those tools in the future, especially those who were already integrating technology in their treatment setting. Despite this apparent motivation to increase technology, the reality encountered in the clinical setting is that a minority of therapists currently include it in their practice [23]. Many factors may contribute to this reluctance. First of all, current practice routines are often hard to break [35]. Moreover, these devices are often not covered by health insurance, which limits their adoption in cognitive rehabilitation. The lack of technological and financial supports in the workplace and the time investment needed for professionals to learn how to use AT may hinder their initiative to include them in their treatment methods. This is especially true in the current financial context of the Canadian healthcare system, where professionals are asked to be more efficient with less resources [31]. Finally, a lack of experience, comfort and knowledge about ATCs has a negative influence on the attitude of professionals towards ATCs, and consequently their use. This last factor is especially important to consider as clinician knowledge is one of the key factors that determine whether consumers are appropriately matched to assistive technologies.

Also, these professionals work in a complex system involving administrators and an organizational context [36]. For example, in this context if a professional needs training to use an ATC, administrators must offer support through monetary compensation and time off. Like all other changes in practice, to succeed organizational barriers must be considered [35].

2.2.2 Patients and Caregivers

As mentioned at the beginning of this chapter, cognitive disabilities such as difficulty conceptualizing, planning, sequencing thoughts/actions and remembering can lead to barriers in performing daily activities at home, at work or in the community. Assistive technology can reduce the effect of these disabilities and improve quality of life [10]. However, before using technology with people with cognitive disorders, it is necessary to understand what their needs are and what barriers limit their use of technologies. According to a study by Kaye [37], persons with disabilities are much less often in contact with computers compared to persons without disabilities, thus contributing to their unfamiliarity with technology. Moreover, a large percentage of those who own devices either do not use them at all or do not use them effectively [10]. The underutilization of ATCs by patients can be explained either by a mismatch between owners' needs and the technology, including inappropriate needs assessments and poor device selection. It has to be considered that users suffering from cognitive disorders often have problems identifying their own needs, which makes it difficult for health professionals to conduct a complete and reliable assessment. Also, psychological factors, which include unrealistic expectations regarding the technology, lack of awareness of one's limitations, lack of support from caregivers, lack of training and lack of information about the benefits of ATCs, often play a role in the poor utilization of ATCs [10, 23]. Indeed, ongoing training and support from professionals is considered vital to the success of compensatory devices [38]. The cost of the technology also has to be taken into account in the list of obstacles that hinder the use of technology by patients [23]. However, although cost is often cited as a barrier to procuring and using technology, users' cognitive and physical impairments add to the challenge [39]. Persons suffering from cognitive impairments are very sensitive to changes in their routines and environment. Finally, the results of a study by de Joode et al. [23] suggest that caregivers sometimes consider the devices as being unsuitable for patients. Also, they often feel they are not properly equipped to support patients with using assistive devices at home.

These challenges were identified by studies conducted many years ago. Since research on assistive devices is currently expanding and technological developments increase their ease of use, portability and intelligence, we can expect that the number of interventions based on ATCs will grow in the coming years. This is especially true at a time when the social trend is to use more and more technology on a daily basis. However, it is important to remember that ATC devices are not necessarily appropriate for all individuals with cognitive deficits. This determination must result from a careful assessment of the person's needs.

2.3 Issues Related to the Commercialization and Marketing of Assistive Technologies for Cognition

The last category of issues that need to be discussed is related to the commercialization of ATCs. These issues are also numerous and they represent another barrier to the widespread adoption of assistive technologies (ATs).

First of all, before purchasing an ATC, users or, in some instances, their caregivers have to be convinced that a specific device will meet their needs. Investors also have to believe in a specific product before investing large sums of money to promote its development and deployment. The effectiveness of new ATCs or ATs must thus be demonstrated to those parties. To do so, rigorous studies with good methodological quality must be conducted to evaluate the utility and impact of new ATCs or ATs in the user's real life and over an extended period of time. Unfortunately, most of the studies that evaluated the efficacy of ATCs are of poor methodological quality; for example, they were conducted without a control group or with very few participants [8, 25, 40]. The lack of valid and strong research evidence about ATCs in a real-life context makes it harder for healthcare professionals to promote their use. It can also reduce potential investors' attraction to those devices. Private parties are often skeptical about the functionalities offered by prototypes and only the few that show solid results get through the next phase. Furthermore, there is a lack of studies showing how to successfully implement an ATC in a clinical setting.

Second, it is a real challenge to measure the cost-effectiveness for the healthcare system of using ATCs, and thus the monetary gains related to their use. The effects, direct and indirect, can go beyond a calculation of costs related to their use. For example, improvement in the person's quality of life has been reported [23, 25], as well as improvement in self-esteem, overall satisfaction and emotional stability [41]. Also, the impact of ATCs on reducing caregiver's burden has been reported [25, 41]. For the same reason, it is hard to calculate a good Return On Investment (ROI) or to predict it for the industry. Such a ROI forecast is usually what attracts an investor to provide venture capital. It also means that the business model for ATCs does not necessarily follow a classic product selling paradigm.

Another important consideration is the value of clinical studies in the eyes of private partners or anyone interested in marketing new products. Such studies may hold great interest for scientists but businesspeople generally prefer market research [10]. Additionally, researchers might face more headwinds when trying to partner with relatively large companies. The main reason is that it is often difficult and risky for companies to modify the way they assemble their products. The addition of one simple element can significantly raise production costs or introduce instability leading to lesser quality. To convince them to make the changes required for research or clinical purposes, researchers have to be more concerned about the reality and interest of the industry. Oddly enough, other difficulties may also arise when a company deals directly with the healthcare system, especially if it is public.

Indeed, a public system is a huge machine and changes usually occur over very long periods of time, which can dramatically delay the commercialization of a product.

2.3.1 Marketing in Conjunction with Clients

Currently, many technologies on the market are not designed for people who have cognitive disabilities. While these technologies may sometimes have great potential to assist in daily living, procedures for operating those devices can be complex for a person with cognitive deficits [42, 43]. Any ATC for people with cognitive disabilities must accommodate the individual's skills and deficits. This is complicated by the fact that each person has a unique combination of strengths and weaknesses [10], as well as different expectations, preferences and past experiences with technology. Indeed, aside from their cognitive disorders, these people often have physical and sensory limitations as well. Possible accommodations for people with cognitive disabilities include visual displays with reduced clutter, provision of information in non-text formats (e.g. graphics, video, audio), minimization of the number and complexity of decision-making points, presentation of information sequentially, and reduced reliance on memory [42]. In addition to this complexity, the technology design and prescription also require consideration of all those who will be in contact with the technology, including clinicians and caregivers as well as people with disabilities [44]. Thus, a high degree of customization is often needed for a cognitive device to be effective. Although there are hundreds of millions of people worldwide with cognitive disabilities, the diversity of conditions and situations means that sales volumes for any given product are low and prices are correspondingly high [10]. High prices mean not only that consumers must pay more but also that the growth of the industry is restricted since many potential consumers cannot afford the prices [10].

2.3.2 Insurance

Another major obstacle to the marketing of ATCs is that in most cases medical insurance does not cover their purchase. Consequently, many patients living on fixed incomes are unable to purchase ATCs and/or service contracts and data plans [10, 23]. The current view is that because ATCs (e.g. smart phones, tablets) are not manufactured for the express purpose of compensating for a disability, private insurance companies and Medicare will not fund their purchase [45]. The "who will pay" question is very important but also very complex. One way to overcome these challenges would be to have the product approved by Official organizations like the Food and Drug Administration in the United States. Such a venture would, however, be very difficult and require significant efforts by a research team.

3 AP@LZ: Lessons Learned

The DOMUS laboratory was founded in 2002 with the goal of promoting and developing technological tools to assist persons with limited autonomy and their caregivers. Over the years, many projects have been developed and most of them have been tested with real users suffering from cognitive impairments. The evaluation process includes a usability test, satisfaction poll, evaluation in situ and other means to determine whether the prototype meets the requirements for being released onto the market. However, despite our best efforts and the efforts of many other researchers, the traction of ATCs remains limited.

In this section, we discuss one of the projects we designed with a User-Centered Approach. The evaluation process began in 2010 and is still ongoing. This project, called AP@LZ, originates from clinicians' needs coupled with MOBUS, a prototype previously designed at the laboratory for people with schizophrenia [46]. It falls within the array of various electronic memory aids designed for persons with cognitive impairments [47, 48].

This section is organized in three parts. In the first part, we present an overview of the project and AP@LZ functionalities. In particular, we describe what methods we used to design AP@LZ. In the second part, we outline the various sets of experiments we conducted over the years. The third and last part highlights the lessons we have learned from this project in relation to valorization.

3.1 What Is AP@LZ?

AP@LZ is an acronym that stands for "Agenda Personnalisé pour des personnes avec la maladie d'ALZheimer" (in English: a personalized agenda for persons with Alzheimer's Disease). The first version of AP@LZ ran on a smart phone under a Microsoft system. It evolved to embrace smart phone trends and was then developed on Android to enable more powerful hardware capabilities, such as increased memory, and functionalities. Table 1 presents the development process interspersed with experimental tests.

The development of AP@LZ was driven by the desire to replace a paper agenda that was being used in the healthcare system in Quebec, Canada. The paper agenda compensates for memory losses and deficits in planning but obliges elders with Alzheimer's Disease (AD) to consult it regularly to remember upcoming activities. Furthermore, information entered by elders with AD may not be precise or clear. Thus it is sometimes difficult for them to understand what they have written previously. Clinicians identified these issues and asked for an electronic agenda to be designed that meets elders' with AD needs and is adapted to their abilities. The agenda was designed according to a user-centered approach. A neuropsychologist who had raised the paper agenda drawbacks represented the user. During three preliminary working sessions, the neuropsychologist and computer scientists

Table 1 AP@LZ development and experiments

Development phase	Method	Participants
Version 1		
Participatory design Agile programming	18 meetings	Neuropsychologist
Test in situ	Usage at home for 1 week	2 elders without cognitive impairments
Test in situ	Usage at home for 6–12 months	2 elders with AD
Version 2		
Transfer to new operating system	4 meetings	Neuropsychologist designer
Usability test	2 h in the laboratory	14 elders without cognitive impairments
		8 elders with AD
Test in situ	Usage at home for 2 months	5 elders with AD

discussed the elders' needs and how the new technology could meet them. Based on observations from her practice, the neuropsychologist explained which function-alities and information are essential for elders with dementia to cope with their everyday organization. For their part, the computer scientists presented MOBUS, a mobile agenda that had been designed at the DOMUS laboratory for people with schizophrenia [46, 49]. Some DOMUS functionalities are interesting for people with AD while others are useless or need changes. For instance, MOBUS helps users remember their appointments and activities but does not allow them to enter them by themselves. The activities entered in MOBUS are chosen jointly by the persons with schizophrenia and their caregivers, and entered by the caregivers. As a baseline, a discussion determined what to reproduce from MOBUS and what kind of functionalities were missing in that agenda. It was agreed that the new appli-cation needed to be simpler.

To develop the prototype, the programmers followed an agile programming approach [50]. The goal was to develop and get feedback as often as possible in order to improve the prototype at every step. Over the five months of development, the clinicians and programmers met 18 times. At each meeting, the computer scientists presented some prototype interfaces and a discussion followed about what to improve and what to keep. Some challenges arose. On a small screen, for people with AD who have difficulty focusing on the appropriate information, the first challenge was to select the essential functionalities without overwhelming the screen. Another challenge was to design the agenda application: how to record the many types of information necessary to add activities, how and when to alert the user of an upcoming activity, whether to include complex functions such as managing occurrences over weeks or modifying an appointment already entered? A third major challenge was the depth of information processing. The depth, i.e. the

Fig. 1 Homepage of
AP@LZ, version 1

steps to go through in various screens to execute a specific task, had to be mini-mized so that people with AD would not lose track of the purpose of the task they planned to do when using AP@LZ. Eighteen meetings were needed to ensure that the iterative process resulted in a version that both parties found satisfactory. Figure 1 shows the homepage (in French) of AP@LZ, version 1.

The user-centered design resulted in a highly functional version of the AP@LZ application that could be tested by the end users, people with Alzheimer's. With this version we conducted some experiments, which are described in the next subsection, but many researchers on our team still wanted to improve AP@LZ after these initial experiments. The first reason was changes in mobile technology and AP@LZ needed to be transferred to Android to exploit the full power of the features available in new smart phones (we chose Android for its openness and wide adoption). Smart phones not only evolved to include more sensing tech-nologies, they also acquired more processing power and much bigger screens (higher resolution). This provided the opportunity to make the interfaces more attractive by asking the designers to improve it. It was also an opportunity to test some remaining issues concerning how to present information for complex func-tionalities. For example, we wanted to test the trade-off between showing all the information on a screen needed to enter an activity or displaying the information step-by-step. Another issue was the location and size of the navigation buttons. As the designer team could not solve these questions, we decided to conduct a usability test. The goal was to ensure elders are able to use AP@LZ easily and to determine between different versions of AP@LZ's interface the one they liked best and found easy to use. Twenty-two elders, including eight with AD, tested the AP@LZ interfaces for two hours in our laboratory or at their home. This study led to a final version of AP@LZ that was much more satisfactory and user-friendly (Fig. 2). On

Fig. 2 Screenshots of AP@LZ, version 2

its home page AP@LZ displays time, date, day of the week and planned appointments for that day. Six other functions are accessible from the home page. First, the "Schedule a new appointment" section enables persons with AD to schedule their appointments independently (from a prerecorded system). Second, the "Personal information" section displays, as the name indicates, personal information such as name, photograph, address, phone number, etc. Third, "Medical information" is a section containing the person's medical history and list of medications. Fourth, the "Contacts" page lists the phone numbers of the person's family and friends. Fifth, the "Photos" section allows the user to display a slide show containing some photographs and explanatory text to jog the person's memory. Finally, the "Notepad" section is used to enter any information the user wants to remember.

3.2 Experiments Done with AP@LZ

The AP@LZ application has been the target of many experiments over the years of development and valorization (Table 1). The first experiment was done with AP@LZ version 1. We expected that:

- People with AD would find AP@LZ easy to learn how to use.
- People with AD would use AP@LZ on a daily basis.
- People with AD would improve their daily life organization.

In 2011 [51], the application was pretested with two elders without dementia (or any other abnormal cognitive impairment) for a period of 12–18 days. The goal was to verify ease of use. It was expected that these elders would encounter no difficulties in using it during this period. At the end of this pretest, participants were asked to comment on their experience. Due to the simplicity of the application, we obtained a lot of comments, which is very rare in usability tests with elders. These comments led to several modifications of the application. This experiment also taught us more about the time required to learn the application. The success rate of the participants varied between 73 and 100 % depending on the task. However, as expected, they reported that they would not want to use AP@LZ regularly as it was too simple.

After getting good feedback from the preliminary experiments, a study in situ was conducted with a population with AD. The protocol was two single-case studies, where we compared the capability for each participant to use AP@LZ and improve their autonomy [52]. The project was approved by the Research Ethics Committee of the CSSS Institut Universitaire de Gériatrie de Sherbrooke, and all participants gave their informed consent. The two participants lived at home with a spouse who gave useful feedback on the elder's capabilities. During a learning period that lasted for about ten weeks with two sessions per week, the two participants with AD learn how to use AP@LZ. This learning period was divided into three phases according to the Sohlberg and Mateer learning method [53]: acquisition, application and adaptation. A good success rate is needed to progress to the next phase. When the learning was completed, the two participants used AP@LZ at home for one year. We periodically evaluated if AP@LZ helped them remember appointments better. As the disease evolved during the experiment, we compared these results with other activities that had not been addressed by AP@LZ. Performances on the tasks assisted by AP@LZ (e.g. remembering to take medication) improved or at least stayed stable over time, compared to performances on tasks not assisted by AP@LZ (e.g. remembering the location of keys). The results of the experiment suggested that (1) persons with AD can learn to use new technologies to compensate for their everyday memory problems, and (2) they can use them efficiently in their day-to-day living if the design remains simple. These results also opened up new rehabilitation possibilities for the population with AD. One of the participants liked AP@LZ so much that he was still using it two years later. With these observations, we tried to promote the application over the consumers market. Our conclusions and lessons learned are discussed in the next subsection.

3.3 Lessons Learned with AP@LZ

AP@LZ has been an enlightening project that has occupied an important place at the DOMUS laboratory over the last few years. We have learned many things that will help in the future development of ATCs. While these lessons are based on experience, our field of research requires such experiments to evolve. We first discuss the two lessons derived from our experience in the conception phase:

- ATCs must be designed using a multidisciplinary approach.
- People with special needs must be involved early in the design process.

The next four lessons concern valorization:

- Software-based inventions are difficult to patent.
- The business model must include support and training for using the ATC.
- Partners do not see the commercial advantages of ATCs.
- Commercialization requires expertise.

3.3.1 Conception

First, we must stress how important it is to involve healthcare professionals in the conception process. Clinicians can share their knowledge and experience when trying to pinpoint the needs and abilities of persons with cognitive impairments. They represent the target population when conceiving a new application. This is valued as experiments take time and people with cognitive impairments have difficulty expressing their feelings. Clinicians can also be the link between computer scientists and people with specific needs. The particular users we target are not easy to approach and need to be handled with care. Clinicians can also act as translators and help computer scientists communicate the information in the adequate language for the persons. However, the question remains whether or not direct contact should be established between computer scientists and clients. Also, clinicians can explain the impact of the cognitive impairments on daily situations. For instance, it would be very difficult for computer scientists to imagine the impact of losses in attention. Therefore, without direct observations of persons with cognitive impairments experiencing difficulties, computer scientists do not have a concrete picture of the situations they are supposed to compensate for.

Secondly, we learned the importance of holding focus groups from the start of a project. They help to link the users' needs and perceptions with the application design. However, recruiting persons with AD for this purpose is not necessarily a good idea. They tire very quickly and have difficulty expressing themselves. We prefer to experiment with people with AD when the application becomes stable. Given the cohort effect and the knowledge of formal and informal caregivers, at the beginning it is better to involve elders without impairments and caregivers in the design and evaluation process, as they are close enough to the target population. Their feelings or abilities when using an application are expected to give a good indication of how persons with AD would behave with the application.

3.3.2 Valorization, Commercialization and Marketing

We tried to valorize this project in many ways after the success of our experiments. The first idea we had was to obtain a patent to protect the intellectual property developed and also create value for a future private partner. Many difficulties prevented this enterprise from being successful. First, the patent system is not adapted to the scientific process that is based on disclosure to get grants. Second, AP@LZ is software and software is not supposed to be patentable.

We thus decided to find a private partner without a patent. At first we thought that the standard *product selling* model would be the perfect fit due to the simplicity of AP@LZ, but in fact the experiments made us realize that a learning period and the involvement of healthcare professionals in the process are crucial to its success. Therefore, the business plan must include technological and rehabilitation support.

Another aspect that may limit the interest of private partners is the target population. ATCs are usually designed for a niche market so it is hard to convince partners that the model will be profitable. The second thing that we aimed to do was to put AP@LZ in an online store for health-related products. The advantage was that the framework already existed and it seemed a much simpler way to offer it to the population. However, the store was not ready to make the move toward technology. This shows one of the big problems we faced: we are radically changing ways of thinking and therefore encounter strong resistance. We then decided to contact pharmacies directly. We approached a local franchise, pharmaceutical group and even a pharmaceutical provider. Again we received negative comments about potential profitability. In addition, potential partners expressed a lot of fear regarding our business model; they would have much preferred a model that did not include services.

Last but not least, it is noteworthy that some of our difficulties have to do with the expertise of the laboratory itself. We are a multidisciplinary research team but we do not have particular competencies in doing business. Moreover, valorization requires the investment of a lot of time that must be taken away from time devoted to research and teaching. Finally, from the valorization efforts made for this project, we learned that technological improvement is upsetting the way things work in business and in healthcare. We need to work toward convincing people that it is worth trying.

4 Framework for Valorization

As mentioned earlier, ATCs have the specificities of being both technological and social innovations. Therefore, transitioning successfully from research laboratory to marketplace needs implementation strategies that consider these two aspects and their complexity. For example, it is necessary to have a better understanding of technology promotion and complex challenges related to the implementation of a new rehabilitation practice in the healthcare system [35]. In this section, we present a theoretical model for technology transfer, a framework to help understand the process of implementing new practices in the rehabilitation domain, and finally a reflective tool to facilitate the valorization of ATCs.

4.1   Closer Look at Valorization

The term valorization can vary depending upon the context. In general, technology transfer is an expression widely used in many domains rather than valorization. According Lane [54], the valorization process of a technological product like an AT involved three main stakeholders groups in our societies, government, academic researchers, and manufacturers. Moreover, the collaboration between these groups to pass with success from research laboratories to marketplace is a complex process partially due to the different barriers like working method, domain's cultures and values [55]. To optimize the chance of success of this process, Flagg et al. [56] have proposed a conceptual model (The Need to Knowledge model, NtK) composed of three "phases", nine "stages" and nine "gates" (see Table 2).

In this model, Lane puts the knowledge like the core of the valorization process of technology product. He states that the latter consist to "transform knowledge about user's problems from conceptual ideas into knowledge embodied as technology-based solutions" [56]. According to the NtK model, this transformation is composed of three consecutive phases:

- First, there is a "discovery phase", which is the step a research activity (ex: literature review, brain storming, etc.) conducted to find a solution to the user' problematic. The researcher or team of research for example try to identify a new concept or technology that already exists and match it to the user needs. For example, DOMUS team worked in this type of phase to identify the concept to create an electronic organizer to help people with AD to manage their daily activities using a mobile phone.
- Secondly, there is an "invention phase", which represents the period of time when engineering methods have to use to develop a functional prototype of the results of discovery phase to allow the demonstration of the feasibility of the concept.

Table 2   The Need to Knowledge model [56]

Phases	Stages and gates
Discovery	Stage 1: Define problem and solution
	Stage 2: Scoping
	Stage 3: Conduct research and generate discoveries = discovery output
Communicate discovery state knowledge	
Invention	Stage 4: Build business case and plan for development
	Stage 5: Implement development plan
	Stage 6: Testing and validation = Invention output
Communicate invention state knowledge	
Production	Stage 7: Plan and for production
	Stage 8: Launch device or service = Innovation output
Communicate innovation state knowledge	
Production	Stage 9: Life-circle review/terminate?

- Third, there is a "production phase", which occurs when the functional prototype is ready to be transformed into a marketable product for a mass quantity by performing a battery of tests (e.g.: quality control) and refining the design of the prototype (e.g.: marketing service) in order to have a final product matching with the objectives of the industrial actor.

For Lane, it's necessary to guide the valorization process with efficacy and efficiently due to the different stakeholders involved. Three "stages" and three "decisions gates" composed the phases (discovery, invention and production) to support the progression in the transformation process of an idea into a marketable product. The stages are like special elements that allow the stakeholders to determine the project needs to realize. The decisions gates are similar to checkpoint that permit to verify according the state of the project if it possible to go to the next stage.

4.1.1 Implementing Changes in Rehabilitation

The valorization process is complex. Various actors are involved including academic researchers, students, engineers and healthcare professionals. To implement ATCs in the healthcare system also means changing the practice of professionals. Changing established behavior is very difficult because of the complex relationships between the healthcare system, professionals, patients and carers [35]. However, to foster the process of valorization or maximize the success of ATC use by users (people with disabilities, carers, healthcare professionals), we need to understand how to change practices in the rehabilitation field.

Chaplin [35] developed a model (see Table 3) to improve patient care through changing healthcare professionals' and managers' practices. He suggested that the development of a successful strategy for change is based on an understanding of the types of barriers occurring in healthcare, and different ways to overcome them.

Table 3 Change practice model in the rehabilitation domain [35]

Barriers faced in healthcare	Strategies to change
• Motivation	• Educational materials
• Awareness and knowledge	• Meetings
• Acceptance and beliefs	• Clinical audit and feedback
• Skills	• Outreach visits
• Practicalities	• Patient-mediated strategies
	• Reminder systems
	• Opinion leader

According to NICE [35] there are six types of barriers in the process of adoption of the innovation by healthcare professionals. These barriers are:

1. Awareness and knowledge—Identify what needs to change and why
2. Motivation—External and internal factors involved in changes
3. Acceptance and beliefs—Attitudes and beliefs which have an influence on the behavior
4. Skills—Feeling about abilities to use
5. Practicalities—Availability of structures, processes, facilities, equipment and human resources
6. Out-of-control barriers—Societal or social variables such as public policy or organization structure.

The NICE [35] model contains six ways to overcome these barriers: (1) Educational materials (e.g. booklets, journals, CDs, videos and DVDs, online tools, computer programs) and meetings (e.g. conferences, workshops, training courses and lectures) used in combination are effective in changing behaviors; (2) Educational outreach visits or academic detailing according to needs; (3) Exploiting opinion leaders to motivate and inspire healthcare professionals to achieve the best possible care for patients and to foster the dissemination of information; (4) Employing clinical audit and feedback to collect data about individual or organizational practice to improve quality; (5) Use of reminder systems to provide healthcare professionals with specific information when they need it (e.g. during a patient's consultation); (6) Use of patient-mediated strategies to focus on giving information to patients and the wider public. The valorization process must consider all of these barriers to adoption of the ATCs and should also consider using these ways to overcome barriers [35].

4.1.2 Aspects to Take into Consideration for Valorization

In order to valorize assistive technologies for cognition, many important aspects must be considered. Some of the most important are:

- **Research team's goals**: Does the team wish to create a technology (academic point of view)? A product (business point of view)?
- **Technology targets**: What kind of people are addressed by the technology? Are there multiple users for the ATC?
- **Business opportunity**: What kind of market is targeted? Are there possibilities to broaden the targeted end users? Is packaging possible?
- **Partnerships**: What types of collaboration are involved in the project (all private, public-private collaboration)? Who are the end users to convince?
- **Involvement of healthcare system**: What are the impacts associated with the use of ATCs on healthcare professionals' practices? How can practices be changed to foster the use of ATCs in clinical settings?

To answer these questions, we propose a guide to help research teams optimize the valorization process of their ATCs. We assume that research teams want to introduce their technology into the marketplace to address unmet needs of people with disabilities.

4.2 Proposed Tool: Stop and Think Prior to Creating ATCs

In this section, we propose a simple guide called "Stop and think prior to creating ATCs" that aims to help researchers move toward the valorization of the ATCs they develop. We chose the 'Stop and Think' expression because it is a cognitive strategy from Etscheidt [57] for changing ways of thinking: stop and think before any action. We decided to work on such a tool after finding that the difficulty with diffusion and marketing of ATCs was generalized [23]. Many problems occur at the implementation stage, one of the most important steps to prove the value created by the ATCs. It is often the consequence of not very rigorous procedures from a scientific perspective: inconsistency in the terminology used, lack of details regarding the strategies, missing theoretical frame of reference, etc. [58]. The current consensus in the literature is that it is a good idea to base the valorization of technologies on knowledge transfer models to increase the chances of success [56]. Other authors suggest that to ensure the success of the enterprise, the implementation must begin with good planning right from the start of the project [23, 58].

This tool, shown on Fig. 3, is designed to be used even before having a first working prototype. It lays the groundwork for seven key questions related to the valorization planning process, based on the literature and the experience of the DOMUS laboratory team. It also aims to provide potential solutions to these key questions. Each of the following subsections addresses one of these questions.

4.2.1 Who Will the End Users of This ATC Be?

To foster the implementation of technological innovations, researchers need to plan the procedure carefully. The end users and clients who will use the ATCs in the future need to be considered and personalized strategies designed accordingly [23, 33]. The particularity of ATCs is that the target users are generally varied and involved in the project to different degrees. They are the patients, the healthcare organization, the caregivers and the managers (directors, etc.). It is important to understand who are the clients and make them the priority. With ATCs, it is also important to understand that the project may involve the managers, political leaders and often private companies.

Another important consideration concerns the objective of developing an ATC. If the goal of the research project is to introduce the technology into the marketplace, the manufacturers may be considered the real customers of the product. Bauer and Lane [59] explains that manufacturers have the capacity to produce,

Stop and think prior to creating ATCs

Questions to ask prior to creating ATC to facilitate valorization:

1. *Who will the final users of this ATC be?*

2. *What are the users' needs?*

3. *How will the research process be conducted?*

4. *Where will the ATC be in a few years?*

5. *What will the business model be?*

6. *How can you make your ATC more attractive?*

7. *Which valorisation model best fit the context?*

The guide and this tools are available on the Domus web site: http://www.domus.usherbrooke.ca/

Fig. 3 The "Stop and think prior to creating ATC" guide

distribute, and support products in the marketplace, including those designed for use by people with disabilities. In this case, what roles are played by other actors such as research collaborators and healthcare professionals? According to Bauer, they are stakeholders who may influence the research activities and commercialization outcomes. Finally, the research team must address manufacturers' needs while keeping in mind the end users' perspective.

4.2.2 What Are the Users' Needs?

The second question that must be answered is what the users' needs are. One of the intuitive ways to answer it, at least from a researcher's perspective, is to do a literature review. It is even more important when the client is a person with a cognitive deficit. The consequences of cognitive deficits are diverse and ATCs must often be adapted to the specific profiles of the target population. Despite this, we also suggest conducting interviews with the users. Such interviews often give the team a lot of crucial information and also enable a good match between the person and technology.

The families of the end users also play an important role in the project. Since they often act as natural helpers, they might participate in the learning and utilization of the ATC. Their needs must, therefore, be considered. One of the ways to better understand both the needs of the families and the end users is to rely on information from the professionals in clinical settings. These professionals are often a very good intermediary to translate the needs into a language the researchers understand. One of the objectives that all the potential users of an ATC share is support at home for persons with disabilities [60].

Finally, it is also very important to know the needs of industry, which often focuses on mass production and low costs. Thus it is important to keep the technologies simple and not too personalized, to allow, for example, another population to use the same technology.

4.2.3   How Will the Research Process Be Conducted?

To develop ATCs, one of the best approaches is to use participatory action research [61]. This approach involves the users (patients, families, managers, professionals, etc.) in the research process. One of the assumptions of this approach is that future users can provide the project with unique expertise. More specifically, they shed a different light on the needs that the technology is trying to meet. Involving them in the process should guarantee that the technology will be useful and efficient in the future. In a study published in 2007, Landry et al. [33] concluded that the more researchers and practitioners (i.e. end users) invest in continuous collaboration through exchange and discussions, the more the data collected from such research will be used.

A very important point underlined by the work of Frank Lopresti et al. [10] is the impact of undertesting an ATC. ATCs should be tested with real patients with cognitive impairments despite the difficulty of conducting such research (ethics committee, recruitment, etc.). The experiments should also take place in a realistic environment. The environment has a lot of influence and results obtained in a laboratory context could be biased as a result of environmental stress. Moreover, it is a good idea to involve healthcare professionals in teaching and integrating the ATCs in the person's real-life context.

Finally, it is important to be careful about the inclusion and exclusion criteria when recruiting participants. A sample should be representative of the target population. In other words, a person should not be chosen only to facilitate the progress and success of the experiments.

4.2.4   Where Will the ATC Be in a Few Years?

Valorizing an ATC is not a simple task. It is important to establish a clear vision and specific objectives before starting a project. The researchers should take the time to understand and pinpoint their own motivations in the project. They should know where they want to go and set a number of long-term objectives associated with their research [59]. For example, suppose one is working in ambient intelligence. If the subject being worked on cannot yet be implemented in a house, the ultimate goal of the project remains a research issue regarding the conception and development of the idea [62].

4.2.5 What Will the Business Model Be?

When it is established that one of the objectives of the project is to introduce a product to the market, the research team must define a clear idea of the business model. It is important to convince future private partners (e.g. manufacturers, industry). A business model or plan is a written statement that describes and analyzes the business to be launched. Most private collaborators or investors require a business plan before even considering a proposal. Two points need to be taken into consideration by the research team when establishing the business model:

Target market of the ATC:

The cognitive impairment market is generally small and highly fragmented. Frank Lopresti et al. [10] reported that the diversity of conditions and situations means that sales volumes of assistive technology are too low and prices are correspondingly high. With the goal of fostering commercialization of an ATC and convincing private partners, the research team must determine if they wish to target a specific market for people with cognitive impairments, with the limitations associated with it, or find another type of market. According to Frank Lopresti et al. [10], there are three different opportunities to facilitate commercialization of products. First, they think that using mainstream products to develop ATCs can achieve the low prices that come from high volumes. Second, they suggest that ATCs can be developed using mainstream components as a base. Third, new mainstream markets can be created by the industry actors, taking into consideration which particular characteristics are common to people with cognitive impairments.

Method to sell the ATC:

Transforming a research project on technology into a commercial product is a complex process. An ATC is an assistive product to help people with disabilities in their daily lives. Consequently, its availability in the marketplace needs to be supervised by healthcare professionals. Research teams have to think about this process. Who prescribes the ATC? Physicians? Clinicians? Hospital? Next, the marketplace needs to be defined. Would the ATC be sold in a drugstore like medication or in a hospital?

The costs of the ATC also need to be taken into consideration. Mason et al. [63] reported that ATs are costly to purchase or maintain. With the goal of fostering their usage, it is necessary to determine who will pay for the ATC: Government agencies? Private insurance? End users? Other actors? Solving this question may have an influence on the success or failure of the valorization process. ATCs, like other products that include services, need to have a customer service to help end users if necessary; for example, if the ATC breaks down or needs to be updated. Who will take on this role: Manufacturers? Research team? The same questions can be asked about the training period necessary for the end user to master the ATC. Healthcare professionals need to spend time to learn the functionalities of the ATC and then how to help the end user learn those functionalities. The research team with collaborators also needs to think about who will pay for this time spent by healthcare professionals: Manufacturers? Government agencies? Hospital? End users?

4.2.6   How Can You Make Your ATC More Attractive?

If the participatory action research approach is selected for the project, questions must be answered regarding the specific expectations users will have about your ATC. Which technological platform should be exploited (tablet, pc, smart phone, etc.)? It is important to keep in mind the simplicity and ease of human-computer interaction. Marketing has something to say about the design of attractive software.

Landry et al. [33] identified the determinants of the implementation relative to the innovation itself (not considering the context).

1. Relative advantage: the innovation should create a significant improvement
2. Compatibility: there should be a coherence between the characteristics of the innovation and the context
3. Complexity: it should be easy to understand and use
4. Experiments: it might be interesting to provide the opportunity for professionals to use the innovation during a trial period
5. Observability: ease of perceiving the effects of the innovation
6. Adaptability: possibility of adapting to the context
7. Radicality: level of changes brought about by the innovation
8. Multifunctionality: possibility of using the innovation in a different context for a different clientele
9. Legitimacy: relates to the adoption by neighbor organizations.

4.2.7   Which Valorization Model Best Fits the Context?

As mentioned previously, the lack of rigor in the valorization process, the lack of common terminology and lack of a frame of reference guiding the process are often reasons for implementation failures.

Answering the previous questions should help in the choice of a good model specific to the ATC. As there are several models, the context must be well known to achieve a good match. Here are some examples of models found in the literature. As shown in Table 4, they are matched against different contexts of development and ATC implementation.

Table 4 Example of models matching the context

Context of development and implementation	Example of models found in the literature
Development and implementation toward the industry	Transferring R&D knowledge: the key factors affecting knowledge transfer success: Cummings et al. [64]
	A model for technology transfer in practice. Gorschek et al. [65]
Development and implementation in public healthcare and practice changes for the professionals	Ottawa Model of Research Use. Graham and Logan [66]: innovations in knowledge transfer and continuity of care
	PARiHS Framework: Promoting Action on Research Implementation in Health Services Kitson (1997) [67]
	Conceptual Model for Considering the Determinants of Diffusion, Dissemination, and Implementation of Innovations in Health Service Delivery and Organization. Greenhalgh et al. [68]
	Innovation dans les services publics et parapublics à vocation sociale. Landry et al. [33]
	Organizational Transformation Model. Lukas et al. (2007) [69]
Development of ATC toward mainstream products	Need to Knowledge (NtK) Model. Flagg et al. [56]

5   Conclusion

This chapter discussed challenges with regard to the valorization of ATCs. We began by describing the main reasons leading to the failures of valorization. We discussed the issues related to the nature of innovation, the stakeholders and the marketing/commercialization of the ATC. We thus set the scene for how complex the valorization of ATCs is.

The second section described AP@LZ, an intuitive and interactive agenda aimed at replacing a paper agenda. AP@LZ was one of the most important projects at the DOMUS laboratory and the one that was nearest completion and ready for commercialization for many years. The lessons learned with AP@LZ showed us that a research team needs to have a broader vision of the assistive technology project which goes beyond "academic research" to include "business culture".

Finally, we proposed a simple guide for the valorization of ATCs. We discussed the importance of taking the time to think about all the implications of the assistive technology project at the very beginning of the project. The guide reviewed seven important questions that, if answered, should foster the success of valorization. However, due to the complexity of the process, it is highly likely that no guide can ever guarantee the success of such an enterprise.

The DOMUS laboratory is currently leading a large-scale project that exploits the ideas expressed in the guide in this chapter. The use and results will be documented in the future. This should enable us to assess the efficacy of the proposed solutions.

References

1. Ricker, J.H.: Traumatic brain injury rehabilitation: is it worth the cost? App. Neuropsychol. **5**, 184–193 (1998)
2. DePompei, R., Gillette, Y., Goetz, E., Xenopoulos-Oddsson, A., Bryen, D., Dowds, M.: Practical applications for use of PDAs and smartphones with children and adolescents who have traumatic brain injury. NeuroRehabilitation **23**, 487–499 (2008)
3. Pavolini, E., Ranci, C.: Restructuring the welfare state: reforms in long-term care in Western European countries. J. Eur. Soc. Policy **18**, 246–259 (2008)
4. Wimo, A., Prince, M.: World Alzheimer report. Alzheimer's Dis. Int. **21** (2010)
5. A new way of looking at the impact of dementia in Canada. Alzheimer Society Canada (2012)
6. Humphreys, I., Wood, R.L., Phillips, C.J., Macey, S.: The costs of traumatic brain injury: a literature review. ClinicoEcon. Outcomes Res. CEOR **5**, 281 (2013)
7. Nations, U.: World population ageing 2009. United Nations, Department of Economic and Social Affairs, Population Division (2010)
8. Nugent, C., Mulvenna, M., Moelaert, F., Bergvall-Kareborn, B., Meiland, F., Craig, D., Davies, R., Reinersmann, A., Hettinga, M., Andersson, A.-L., Droes, R.-M., Bengtsson, J.E.: Home based assistive technologies for people with mild dementia. In: Proceedings of the 5th International Conference on Smart Homes and Health Telematics. Springer, Nara (2007), pp. 63–69
9. Haskins, E.C., Cicerone, K.D., Trexler, L.E.: Cognitive Rehabilitation Manual: Translating Evidence-Based Recommendations into Practice. ACRM Publishing (2012)
10. Frank Lopresti, E., Mihailidis, A., Kirsch, N.: Assistive technology for cognitive rehabilitation: State of the art. Neuropsychol. Rehab. **14**, 5–39 (2004)
11. Sohlberg, M.M.: Cognitive rehabilitation manual: translating evidence-based recommendations into practice. Arch. Clin. Neuropsychol. **27**, 931–932 (2012)
12. Cook, D.J., Youngblood, M., Heierman, E.O., Gopalratnam, K., Rao, S., Litvin, A., Khawaja, F.: MavHome: an agent-based smart home. In: Proceedings of the First IEEE International Conference on Pervasive Computing and Communications. IEEE Computer Society (2003), pp. 521–524
13. Coyle, L., Neely, S., Rey, G., Stevenson, G., Sullivan, M., Dobson, S., Nixon, P.: Sensor fusion-based middleware for assisted living. In: Proceedings of 1st International Conference on Smart Homes & Heath Telematics (ICOST'2006) "Smart Homes and Beyond". IOS Press (2006), pp. 281–288
14. Bouchard, K., Bouchard, B., Bouzouane, A.: Guideline to efficient smart home design for rapid AI prototyping: a case study. In: International Conference on PErvasive Technologies Related to Assistive Environments. ACM (2012)
15. Giroux, S., Leblanc, T., Bouzouane, A., Bouchard, B., Pigot, H., Bauchet, J.: The praxis of cognitive assistance in smart homes. In: Gottfried, B., Aghajan, H.K. (eds.) Behaviour Monitoring and Interpretation, vol. 3, pp. 183–211. IOS Press, BMI Book (2009)
16. Demerism, G., Hensel, B.K., Skubic, M., Rantz, M.: Senior Residents' Perceived Need of and Preferences for "Smart Home" Sensor Technologies. Cambridge University Press, Cambridge (2008). (ROYAUME-UNI)
17. Augusto, J.C., Nugent, C.D.: Smart homes can be smarter. In: Designing Smart Homes: Role of Artificial Intelligence, vol. 4008. Springer, Berlin (2006), pp. 1–15

18. Wilson, B., Emslie, H., Quirk, K., Evans, J.: Reducing everyday memory and planning problems by means of a paging system: a randomised control crossover study. J. Neurol. Neurosurg. Psychiatry **70**, 477–482 (2001)
19. Gentry, T., Wallace, J., Kvarfordt, C., Lynch, K.B.: Personal digital assistants as cognitive aids for individuals with severe traumatic brain injury: a community-based trial. Brain Inj. **22**, 19–24 (2008)
20. Imbeault, H., Bier, N., Pigot, H., Gagnon, L., Marcotte, N., Fulop, T., Giroux, S.: Development of a personalized electronic organizer for persons with Alzheimer's disease: the AP@ lz. Gerontechnology **9**, 293 (2010)
21. de Joode, E., van Heugten, C., Verhey, F., van Boxtel, M.: Efficacy and usability of assistive technology for patients with cognitive deficits: a systematic review. Clin. Rehab. (2010)
22. Wilson, B.A., Emslie, H., Quirk, K., Evans, J., Watson, P.: A randomized control trial to evaluate a paging system for people with traumatic brain injury. Brain Inj. **19**, 891–894 (2005)
23. de Joode, E.A., van Boxtel, M.P., Verhey, F.R., van Heugten, C.M.: Use of assistive technology in cognitive rehabilitation: exploratory studies of the opinions and expectations of healthcare professionals and potential users. Brain Inj. **26**, 1257–1266 (2012)
24. Pollack, M.E.: Intelligent technology for an aging population: the use of AI to assist elders with cognitive impairment. AI Mag. **26**, 9 (2005)
25. Scherer, M.J., Hart, T., Kirsch, N., Schulthesis, M.: Assistive technologies for cognitive disabilities. In: Critical Reviews™ in Physical and Rehabilitation Medicine, vol. 17 (2005)
26. Sohlberg, M.M., Turkstra, L.S.: Optimizing Cognitive Rehabilitation: Effective Instructional Methods. Guilford Press (2011)
27. Mulgan, G., Tucker, S., Ali, R., Sanders, B.: Social innovation: what it is, why it matters and how it can be accelerated (2007)
28. Bessant, J.: Enabling continuous and discontinuous innovation: learning from the private sector. Public Money Manage. **25**, 35–42 (2005)
29. West, M.A., Borrill, C.S., Dawson, J.F., Brodbeck, F., Shapiro, D.A., Haward, B.: Leadership clarity and team innovation in health care. Leadership Q. **14**, 393–410 (2003)
30. Siau, K., Messersmith, J.: Analyzing ERP implementation at a public university using the innovation strategy model. Int. J. Human-Comput. Interact. **16**, 57–80 (2003)
31. Pennacchia, J.: Exploring the Relationships Between Evidence and Innovation in the Context of Scotland's Social Services (2013)
32. Walker, R.M.: Innovation type and diffusion: an empirical analysis of local government. Public Adm. **84**, 311–335 (2006)
33. Landry, R., Becheikh, N., Amara, N., Halilem, N., Jbilou, J., Mosconi, E., Hammami, H.: Innovation dans les services publics et parapublics à vocation sociale: Rapport de la revue systématique des écrits. Transfert des connaissances et l'innovation. Québec: C FCRSS/IRSC (2007)
34. Duperré, M., Plamondon, A.: Innovations sociales dans les organismes communautaires: facteurs intervenant dans le processus de transfert des connaissances. CRISES (2006)
35. Chaplin, S.: How to change practice—A NICE guide to overcoming the barriers. Prescriber **19**, 47–50 (2008)
36. Caouette, M., Lussier-Desrochers, D.: Comment accompagner l'implantation des technologies de soutien à l'intervention dans les milieux de pratique?
37. Kaye, H.S.: Computer and internet use among people with disabilities. Disability Statistics Report 13 (2000)
38. Scherer, M.J.: The change in emphasis from people to person: introduction to the special issue on assistive technology. Disabil. Rehabil. **24**, 1–4 (2002)
39. Wright, P., Rogers, N., Hall, C., Wilson, B., Evans, J., Emslie, H., Bartram, C.: Comparison of pocket-computer memory aids for people with brain injury. Brain Inj. **15**, 787–800 (2001)
40. Bouchard, B., Bouchard, K., Bouzouane, A.: A smart range helping cognitively-impaired persons cooking. In: Twenty-Sixth Annual Conference on Innovative Applications of Artificial Intelligence. Association for the Advancement of Artificial Intelligence (AAAI) (2014)

41. Bergman, M.M.: The benefits of a cognitive orthotic in brain injury rehabilitation. J. Head Trauma Rehab. **17**, 431–445 (2002)
42. Wehmeyer, M.L.: National survey of the use of assistive technology by adults with mental retardation. Ment. Retard. **36**, 44–51 (1998)
43. Lynch, B.: Historical review of computer-assisted cognitive retraining. J. Head Trauma Rehab. **17**, 446–457 (2002)
44. Cole, E., Ziegmann, M., Wu, Y., Yonker, V., Gustafson, C., Cirwithen, S.: Use of "Therapist-Friendly" tools in cognitive assistive technology and telerehabilitation. In: Proceedings of the RESNA International 2000 Conference, Orlando, Florida, USA, 28 June 2000 (2000), pp. 31–33
45. Golinker, L.: Key questions for medicare coverage & funding for AAC devices (2001)
46. Giroux, S., Pigot, H., Paccoud, B., Pache, D., Stip, E., Sablier, J.: Enhancing a mobile cognitive orthotic: a user-centered design approach. Int. J. Assistive Rob. Mechatron. **9**, 36–47 (2008)
47. Wilson, B.A., Evans, J.J., Emslie, H., Malinek, V.: Evaluation of NeuroPage: a new memory aid. J. Neurol. Neurosurg. Psychiatry **63**, 113–115 (1997)
48. Oriani, M., Moniz-Cook, E., Binetti, G., Zanieri, G., Frisoni, G., Geroldi, C., De Vreese, L., Zanetti, O.: An electronic memory aid to support prospective memory in patients in the early stages of Alzheimer's disease: a pilot study. Aging Mental Health **7**, 22–27 (2003)
49. Sablier, J., Stip, E., Jacquet, P., Giroux, S., Pigot, H., group, M., Franck, N.: Ecological assessments of activities of daily living and personal experiences with Mobus, an assistive technology for cognition: a pilot study in schizophrenia. Assistive Technol. **24**, 67–77 (2012)
50. Beck, K., Beedle, M., Van Bennekum, A., Cockburn, A., Cunningham, W., Fowler, M., Grenning, J., Highsmith, J., Hunt, A., Jeffries, R.: Manifesto for agile software development (2001)
51. Imbeault, H., Pigot, H., Bier, N., Gagnon, L., Marcotte, N., Giroux, S., Fülüp, T.: Interdisciplinary design of an electronic organizer for persons with Alzheimer's disease. In: Toward Useful Services for Elderly and People with Disabilities. Springer, Berlin (2011), pp. 137–144
52. Imbeault, H., Bier, N., Pigot, H., Gagnon, L., Marcotte, N., Fulop, T., Giroux, S.: Electronic organiser and Alzheimer's disease: fact or fiction? Neuropsychol. Rehab. **24**, 71–100 (2014)
53. Sohlberg, M.M., Mateer, C.A.: Training use of compensatory memory books: a three stage behavioral approach. J. Clin. Exp. Neuropsychol. **11**, 871–891 (1989)
54. Lane, J.P.: The state of the science in technology transfer: implications for the field of assistive technology. J. Technol. Transfer **28**, 333–354 (2003)
55. Lane, J.P., Flagg, J.L.: Translating three states of knowledge-discovery, invention, and innovation. Implementation Sci. **5**, 1–14 (2010)
56. Flagg, J.L., Lane, J.P., Lockett, M.M.: Need to Knowledge (NtK) Model: an evidence-based framework for generating technological innovations with socio-economic impacts. Implementation Sci. **8**, 21 (2013)
57. Etscheidt, S.: Reducing aggressive behavior and improving self-control: a cognitive-behavioral training program for behaviorally disordered adolescents. Behav. Disord. (1991)
58. Proctor, E.K., Powell, B.J., McMillen, J.C.: Implementation strategies: recommendations for specifying and reporting. Implementation Sci. **8**, 139 (2013)
59. Bauer, S.M., Lane, J.P.: Convergence of assistive devices and mainstream products: keys to university participation in research, development and commercialization. Technol. Disabil. **18**, 67–77 (2006)
60. Carrier, A., Levasseur, M., Mullins, G.: Accessibility of occupational therapy community services: a legal, ethical, and clinical analysis. Occup. Therapy Health Care **24**, 360–376 (2010)
61. Cave, A., Ramsden, V.: La recherche-action participative [Hypothèse: la Page de la recherche]. Can. Fam. Physician **48**, 1671 (2002)
62. Lane, J.P.: Understanding technology transfer. Assistive Technol. **11**, 5–19 (1999)

63. Mason, S., Craig, D., O'Neill, S., Donnelly, M., Nugent, C.: Electronic reminding technology for cognitive impairment. Brit. J. Nurs. **21**, 855–861 (2012)
64. Cummings, J.L., Teng, B.-S.: Transferring R&D knowledge: the key factors affecting knowledge transfer success. J. Eng. Tech. Manage. **20**, 39–68 (2003)
65. Gorschek, T., Wohlin, C., Carre, P., Larsson, S.: A model for technology transfer in practice. IEEE Softw **23**, 88–95 (2006)
66. Graham, I.D., Logan, J.: Innovations in knowledge transfer and continuity of care. CJNR (Can. J. Nurs. Res.) **36**, 89–103 (2004)
67. Rycroft-Malone, J.: The PARIHS framework—A framework for guiding the implementation of evidence-based practice. J. Nurs. Care Qual. **19**, 297–304 (2004)
68. Greenhalgh, T., Robert, G., Macfarlane, F., Bate, P., Kyriakidou, O.: Diffusion of innovations in service organizations: systematic review and recommendations. Milbank Q. **82**, 581–629 (2004)
69. Lukas, C.V., Holmes, S.K., Cohen, A.B., Restuccia, J., Cramer, I.E., Shwartz, M., Charns, M. P.: Transformational change in health care systems: an organizational model. Health Care Manage. Rev. **32**, 309–320 (2007)

Authors Biography

Stéphanie Pinard is a Ph.D. candidate at the School of Rehabilitation of the University of Montreal. She received her bachelor's degree in occupational therapy from the University of Montreal and a master's degree in rehabilitation from the University of Sherbrooke. She is involved in 6 different courses in the master's programs at the School of Rehabilitation of the University of Sherbrooke. Since 2003, she has worked in a rehabilitation center as an occupational therapist with individuals with cognitive impairments.

Kevin Bouchard is a postdoctoral fellow at the University of Sherbrooke. He obtained his Ph.D. from the University of Quebec at Chicoutimi (UQAC) in 2014. His research, primarily focused on activity recognition in smart homes, spatial reasoning and learning, has earned him many prestigious scholarships, including the Alexander Graham Bell Canada Graduate Scholarship from the Natural Sciences and Engineering Research Council of Canada (NSERC).

Yannick Adelise is a Ph.D. candidate in computer science at the University of Sherbrooke. He received his bachelor's degree in computer science from the University of Bordeaux (France) and two master's degrees, one in cognitive sciences and one in neurosciences, from the University of Bordeaux.

Véronique Fortin is an occupational therapist at the Centre de réadaptation Estrie who also works as a research assistant at the DOMUS laboratory at the University of Sherbrooke. She received her master's degree in rehabilitation from the University of Sherbrooke in 2011.

Hélène Pigot is a Professor at the University of Sherbrooke and one of the founders of the DOMUS laboratory. She obtained her Ph.D. in computer science from the University of Paris VI. Her research interests encompass various areas such as cognitive assistance, smart homes, ubiquitous computing, activity recognition and human-computer interaction.

Nathalie Bier is an assistant professor at the School of Rehabilitation of the University of Montreal. She received her bachelor's degree in occupational therapy and a master's degree in rehabilitation at the University of Montreal. She obtained her Ph.D. in gerontology from the University of Sherbrooke. Her research interests include cognitive assistive technologies, smart homes, assessment of activities of daily living, and cognitive rehabilitation in different forms of dementia.

Sylvain Giroux is a professor at the University of Sherbrooke. He obtained his Ph.D. in computer science from the University of Montreal in 1993. He is currently the director of the DOMUS laboratory. His research interests include assistive technologies, smart homes, context awareness and ubiquitous computing.

Safe and Automatic Addition of Fault Tolerance for Smart Homes Dedicated to People with Disabilities

Sébastien Guillet, Bruno Bouchard and Abdenour Bouzouane

Abstract In this chapter, we discuss a project of the LIARA laboratory that introduces a methodology to design and control smart homes dedicated to people with disabilities. In this context, this project aims at improving the security of the environment through a design methodology involving formal synthesis techniques.

Keywords Smart home · Security · Fault tolerance · Discrete controller synthesis

1 Introduction

Ubiquitous computing, making us more connected to our environment and other people, is challenging the way we live through many different means, ranging from anticipating our needs to securing our environment and automating routine physical tasks. Contributions to ubiquitous computing has lead the scientific community to the smart home era [1], which involves a wide range of these means to liberate us from usually hard and repetitive work at home and to help us live more independently.

Enhancing independence is actually the core concept of smart homes dedicated to disabled people. For example, such a house can be designed to help a human resident suffering from a cognitive deficit to complete his activities of daily living (ADL) [2] without the need of additional human assistance. De-signing this kind of smart homes involves many challenges, including blending unobtrusively into the home environment [3], recognizing the ongoing inhabitant activity [4], localizing

S. Guillet (✉) · B. Bouchard · A. Bouzouane
LIARA, Université du Québec À Chicoutimi, 555 Blvd Université EST, G7H2B1
Chicoutimi, QC, Canada
e-mail: sebastien.guillet1@uqac.ca
URL: http://liara.uqac.ca

B. Bouchard
e-mail: bruno.bouchard@uqac.ca

A. Bouzouane
e-mail: abdenour.bouzouane@uqac.ca

© Springer International Publishing Switzerland 2016
K.K. Ravulakollu et al. (eds.), *Trends in Ambient Intelligent Systems*,
Studies in Computational Intelligence 633, DOI 10.1007/978-3-319-30184-6_4

objects [5], adapting assistance to the person's cognitive deficit [6], and securing the environment [7].

Given the high degree of vulnerability of people with cognitive deficiencies, securing the house is a primary concern. Indeed, an adequately designed smart home for disabled people should be able to provide both assistance and protection. However, even if a smart home system is usually build to last, it might not be the case for its very own components [8]: lights, screens, sound system, and many other important equipment can fail during the lifetime of the system. In this context, providing a viable security strategy over time requires to take failures into account due to their high probability and potential harmful consequences if not taken seriously [9].

To operate properly over time, the main concern of a fault tolerant smart home system is, upon detection of a failure, knowing how to "react appropriately". Let's suppose that a detected random failure affects an arbitrary component, is the system still able to provide both protection and adequate assistance with respect to the person's disabilities?

Answering such a question usually requires to solve a non-trivial combinatorial problem: a smart home is supposed to be composed of many dynamical components (electrical shutters, lights, ventilation systems, etc.), each one having several exclusive execution modes (opening, opened, on, off, disabled, failed, etc.) which can be observed using sensors; These components are concurrently executed and their execution modes can be influenced upon reception of events which can be external to the system (e.g. the user pushes a button) and/or internal (e.g. a security system prevents a hair dryer from powering on because it is too hot). Here lies the complexity: ensuring that the system will respect a security property (e.g. being able to provide assistance even if a component fails) requires to verify that this property holds for each accessible combination of execution modes.

Now let's introduce the notion of *controllability*, which happens when a component offers an interface so that a control system—a program named controller—can send events to constrain its behaviour. Controllability is very common in the context of ubiquitous computing—smart home is no exception—where almost every component provides such an interface so that it can be adapted to a situation. A system is said to be controllable with respect to a temporal property whether given its dynamicity and controllability, it exists a controller able to constrain the system such that the temporal property holds for all possible executions.

Applied to smart homes for disabled people, a smart home has the capacity to undertake a security constraint[1] over time if and only if a control system can be proven to keep the system in execution modes complying with the constraint. But even if a system under control can be proven correct using verification techniques, its controller is not guaranteed to be interesting. For example, let's take a controllable component which can be prevented to start, and a security constraint such that this component must not be started when temperature is above 50 °C; now let's

---

[1]Or, to be more general, a "quality of service constraint".

build a controller which always disables this component; then the system under control can be verified to be correct, however we understand that the implemented controller should be more *permissive* when temperature is lower than 50 °C.

Basically, in presence of controllability, the designer of a fault tolerant smart home system has to face two non-trivial problems: building a permissive controller and verifying the system under control. It happens that these two problems are the specialty of a formal technique named Discrete Controller Synthesis (DCS) [10]: given a system's dynamicity, controllability, and temporal constraints, if a control solution exists then DCS is able to provide automatically a controller which is both correct by construction and maximally permissive, meaning that it valuates control events tied to control interfaces of dynamic components only when the system has to be constrained.

In [11], we made a contribution giving concepts on representing the behavior of a smart home system dedicated to disabled people. DCS was shown to be applicable using this representation to create a controller designed to keep the smart home in a correct state. The contribution of the present study relies on the definition of a design methodology around these concepts, and shows examples on how a smart home system can be specified, so that DCS can be applied to solve concrete fault tolerance related problems in smart homes for people with impairments.

The paper is organized as follows. Section 2 presents the related work about security in the smart home domain and justifies the choice of DCS over other formal techniques to provide a solution for the controllability problem. Section 3 describes the synchronous framework which serves as a foundation for modeling and applying DCS. Section 4 explains how to de ne a smart home model using this synchronous framework. Section 5 shows the application of DCS on such a model. Section 6 details the experiments we conducted using a partial model of our own smart home equipment. The model is kept partial so that both DCS application and controller execution remain easy to follow step by step. Finally, Sect. 7 concludes the paper and outlines future developments of this work.

2 Related Work

The literature on which this study is based can be divided into three major domains. The first one is smart home modeling [7, 12, 13]: this work aims to give a framework to represent key aspects of a smart home (*dynamicity*, *controllability*, and *temporal constraints*) so that formal techniques such as DCS can be applied; these aspects are generic and need concrete definitions for the smart home context. The second domain is smart home security [2, 6, 9, 14–18]: what makes a smart home secure, especially a smart home for disabled people? What are the techniques employed to provide some form of security in this context? Finally, the third domain is formal techniques [11, 19–22]: failing to provide a correct smart home behavior for all its possible executions could easily have harmful consequence for a vulnerable inhabitant, so how do we prove a smart home to be secure? And what is the best technique to apply given the smart home properties?

2.1 Smart Home Modeling

Research projects related to modeling of smart homes for disabled people usually share many concepts based on representing: the smart home elements (devices, doors, lights, etc.) with their positions and execution modes, the person itself (its state of mind, behavior, position, cognitive profile, etc.), and the global execution model (how the smart home is supposed to run and process events in order to provide both assistance and security using artificial intelligence).

In [7, 12], Pigot et al. present respectively (1) a meta-model containing generic knowledge of a smart home system for elders suffering from dementia and (2) a corresponding model showing cognitive assistance and telemonitoring concepts. These works detail a pervasive infrastructure and applications to provide assistance to elders with cognitive deficiencies using two kinds of interventions: one operating inside the home to help the person to complete its ADL in case of difficulties, and another one establishing communication outside the home to send message to caregivers, medical teams or families.

In [13], Lat et al. give an overview of an ontology-based model of a smart home dedicated to elderly in loss of cognitive autonomy. The ontological architecture is partitioned into seven sub-domains: (1) Habitat, describing the home structure (rooms, doors, windows, etc.); (2) Person, which can describe the patient itself (medical history, behavior, etc.) and the various persons supposed to interact with the patient and/or the habitat (medical actor, habitat-staff, friend, etc.); (3) Equipment, which defines the various home appliances; (4) Software, de-scribing reusable software modules of the smart home system; (5) Task, detailing the observable tasks that the patient, the personal, and house itself, can per-form; (6) Behavior, regrouping life habits and critical physiological parameters; (7) Decision, related to the smart home adaptation behavior.

These works constitute the foundation of Sect. 4, which will synthesize and show how to represent their ideas into a formal synchronous model, so that security properties can be set and verified.

2.2 Smart Home Security

Due to its importance, security in the context of smart homes—especially those dedicated to disabled people—has been widely covered in the literature. Methods employed to secure a smart home target three main layers: (1) Fault tolerance, as a smart home system is supposed to experience failures through its lifetime; (2) Smart sensing technology, so that ADL can be monitored accurately; (3) Appropriate smart home behavior depending on the patient deficiencies.

Fault tolerance Failures in a smart home system may occur on several levels [9, 23]: a smart home is typically a set of hardware and software components communicating together, so failures can happen either happen at hardware, software or communication level.

Sensors, actuators, displays, speakers, lights, etc. are traditional failure-prone smart home hardware components. They wear over time, can be damaged, can go down if they are battery powered, can cease to communicate because of limited signal strength, can operate incorrectly because of a manufacturing defect, etc. A single failure at this level can compromise the smart home security.

A smart home system also typically contains multiple software components running together (operating systems, artificial intelligences, controllers, etc.), including commercial applications (i.e. trusted black boxes). Unless formally checked against security requirements, few assumption should be made about applications. Even software verified by competent and credible experts can contain bugs. The malfunction in the control software in Ariane 5 Flight 501 is an example of such a bug, which remained undetected through several human-driven verification processes [24].

Communication between hardware and software components happens through wired and wireless channels. Communication failures are mainly caused by low signal strength (e.g. two mobile wireless devices communicating together get separated by a too long distance) or heavy traffic. They are not really hardware or software related, but can be (wrongly) perceived as such because affected components cease to communicate and become unavailable, making these failures important to detect.

When a hardware, software or communication component failure is detected, two common responses are (1) using an equivalent component (redundancy) [17] and (2) executing the system in a degraded mode, allowing it to work correctly through failures using a safe subset of its functionalities [25].

These methods will be used in Sects. 4 and 5 as a base to build a fault tolerant smart home.

Smart sensing Increasing a smart home robustness also involves an effective sensing system. Identifying ADL [4, 15], locating a person or mobile components [5], recognizing the mood [26], etc. are examples of smart sensing features that can be integrated into a smart home.

Section 4 takes the presence of these kinds of high level sensors (artificial intelligences) into account, so that security rules can be based on their information.

In [17], Bouchard et al. give guidelines to integrate and execute artificial intelligence modules into a generic smart home system. We will take advantage of these guidelines to model a system that will comply with the same execution principles.

Impairment adaptation Knowing how to adapt a smart home for disabled people to their impairments is a sensitive and complex problem largely discussed in the literature. Smart homes usually contain technological devices aiming to provide adapted cognitive assistance—or *prompts*—when needed. Typical prompts can be based on sounds, music, spoken messages, photos, videos, lights, etc. Implementing an adequate prompting system is actually the core concept or impairment adaptation [6, 7, 15, 16].

In [6, 17], the authors provide experimental results on prompt efficiency according to cognitive pro les. Section 4 shows how to represent these relations, so that security properties can be defied for the prompting system.

Combined together, all these security layers bring a new question. If the house is equipped with redundant critical equipment, if the prompting system can be adapted in accordance with the severity and characteristics of the patient's impairments, and if the context (ADL, mood, position, etc.) can be accurately monitored: how do we prove that, in case of a failure, the smart home system can still provide adequate assistance if the failure impacts the prompting system or the way ADL can be monitored?

Proving it for every allowed failure, every possible execution, every context, etc. is essentially a combinatorial explosion problem that is very di cult to solve without appropriate tools. This is where verification techniques come in.

2.3 Formal Methods

Many research work contribute to formal modeling and verification of user's in-teractions, hardware/software components and control algorithms in the smart home domain [27–29]. However, formal verification suppose that a complete system can be modeled before being applied. In the modeling methodology pro-posed in [29], a modeling step named "control algorithm modeling" is explicitly required. This step is about the definition of a module which, given (1) the system current configuration, (2) incoming message from the system or its environment, and (3) control rules, makes a reconfiguration decision and sends triggering mes-sages to the associated devices for performing the required operations.

This step is precisely the part that is di cult to design because of the combina-torial problem we are facing in this context. This is the reason why we are more interested into an alternative method, DCS, which is able to both build the control part automatically and perform formal verification of the system.

Regarding smart homes, a smart home system can be considered as a special-ization of autonomic computing systems [30], which adapt and reconfigure them-selves through the presence of a feedback loop. This loop takes inputs from the environment (e.g. sensors), updates a representation (e.g. Petri nets, automata) of the system under control, and decides to reconfigure the system if necessary. This consideration is detailed in Sect. 3. Describing such a feed-back loop can be done in terms of a DCS problem. It consists in considering on the one hand, the set of possible behavior of a discrete event system [31], where variables are partitioned into uncontrollable and controllable ones. The uncontrollable variables typically come from the system's environment (i.e. "inputs"), while the values of the controllable variables are given by the synthesized controller itself. On the other hand, it requires a specification of a control objective: a property typically concerning reachability or invariance of a state space subset. Such a programming makes use of reconfiguration policy by logical contract. Namely, specifications with contracts amount to specify declaratively the control objective, and to have an automaton describing possible behavior, rather than writing down the complete correct control solution. The basic case is that of contracts on logical properties i.e., involving only Boolean conditions on states and events. Within the synchronous approach [19], DCS has been defined and implemented as a tool integrated with synchronous languages: SIGALI [20]. It handles transition systems with the multi-event labels typical of the synchronous approach, and features weight functions mechanisms to introduce some quantitative information and perform optimal DCS.

One of the synchronous languages it has been integrated with is BZR [22], which is used in this work; BZR actually includes a DCS usage from Sigali within its compilation. The compilation yields (if it exists) the code of a correct-by-construction controller (here in C language), which can itself be compiled to be executed into the smart home system.

Based on the synchronous characteristics of a smart home system, Sect. 3 sets the synchronous context and notations so that they can be applied to smart home modeling in order to perform DCS.

DCS has already been successfully applied in various domains, e.g. adaptive resource management [32], reconfigurable component-based systems [33], reconfigurable embedded systems [11], etc. However, DCS in the context of smart homes has not been seen until very recently, where it was introduced in [34] and [35]. Both studies show preliminary results on how DCS could be applied to secure a smart home, [34] having a general point of view, and [35] a specific one regarding fault tolerance. They have a common perspective to show results with more types of objectives and adaptive control in order to go beyond a demonstration of DCS applicability and really show its relevance and efficiency in this context. This study takes this perspective into account to give a contribution on actual usage of DCS to solve concrete smart home problems—related to fault tolerance—through a modeling methodology using BZR.

3 Synchronous Framework: Basic Notions

Synchronous languages are optimized for programming reactive systems, i.e. systems that react to external events. This section aims at presenting the similarities between a reactive system under control and a controlled smart home, so that a synchronous framework—essentially adopted from [36, 37]—gets justified as appropriate to specify smart home systems.

3.1 Execution Model

In [11], the execution model of a reactive system under control is depicted, cf. Fig. 1. Such a system contains a global execution loop, which starts by taking events from the environment. Then these events get processed by a task (*Reconfiguration controller*), which chooses the system's configuration. Finally, this configuration order gets dispatched through the system's tasks following its model of computation, and another iteration of the loop can start again. If a system can be represented within this execution model, then the proposition of this work can help to design and formally obtain its *Reconfiguration controller* task.

In [17], guidelines to build the software architecture of a smart home system are presented, cf. Fig. 2. Such a software follows a loop-based execution, in which a database containing an updated system state and event values is read and processed by eventual artificial intelligence (AI) modules to transform raw data into high level information. This information can then be used by third party applications.

Immediately, we can see similarities arising from such an architecture com-pared to the reactive system execution model. If we add a reconfiguration controller as a third party application in this software architecture, then we obtain the same execution principle presented in Fig. 1: in each iteration of the execution loop a controller can be designed to (1) take events and/or high level information provided by the system and its environment, (2) perform a reconfiguration decision, and

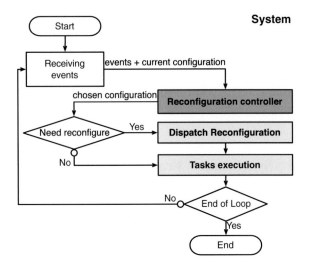

Fig. 1 Configuration processing flowchart

Fig. 2 Smart home software
architecture

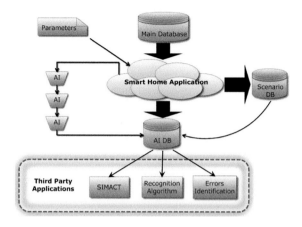

(3) give this decision back to the system using some of its actuators (i.e. its controllability) before the next iteration.

Designing the aforementioned controller by constraint so that it can be obtained automatically through DCS becomes possible, but it requires the use of formal a model to specify the behavior of the underlying system under control. Behavioural modeling can be performed using various formal representations, e.g. State charts, Petri-nets, Communicating Sequential Processes or other ways. The toolset we use in this work—BZR and SIGALI—brings us to de ne our system in terms of synchronous equations and Labelled Transition Systems.

3.2 Synchronous Equation

In a declarative synchronous language, semantic is expressed in terms of data flows: values carried in discrete time are considered as infinite sequence of values, or *flows*. At each discrete instant, the relation between input and output values is defined by an equational representation between flows, it is basically a system of equations: equations are evaluated concurrently in the same instant and not in sequence, the real evaluation order being determined at compile-time from their interdependencies. For example, let x and y be two data flows such that $x = x_0, x_1, \ldots$ and $y = y_0, y_1, \ldots$ Evolution of y over time is given by the following system of equations:

$$\begin{cases} y_0 = x_0 \\ y_t = y_{t-1} + x_t & \text{if } t \geq 1 \end{cases}$$

In this example, y is defined, amongst others, by a reference to its value at a previous discrete instant. Each declarative synchronous language has a syntax to de ne such a system. The corresponding BZR program is: $y = x \rightarrow \text{pre}(y) + x$; meaning that in the first step, y takes the current value of x, and for all next steps

y will take its previous value incremented by x. (Other syntactic features of BZR can be found online[2]). To represent the system execution modes, BZR also allows to define automata, or Labelled Transition Systems, each state encapsulating a set of synchronous equations evaluated only when the state is activated.

3.3 Labelled Transition System (LTS)

A LTS is a structure $S = \langle \mathcal{Q}, q_0, \mathcal{I}, \mathcal{O}, \mathcal{T} \rangle$ where $\mathcal{Q}$ is a finite set of states, q_0 is the initial state of S, $\mathcal{I}$ is a finite set of input events (produced by the environment), $\mathcal{O}$ is a finite set of output events (emitted towards the environment), and $\mathcal{T}$ is the transition relation, that is a subset of $\mathcal{Q} \times \text{Bool}(\mathcal{I}) \times \mathcal{O}^* \times \mathcal{Q}$, where $\text{Bool}(\mathcal{I})$ is the set of boolean expressions of $\mathcal{I}$. If we denote by $\mathcal{B}$ the set $\{\text{true, false}\}$, then a guard g belong to $g \in \text{Bool}(\mathcal{I})$ can be equivalently seen as a function from $2^{\mathcal{I}}$ into $\mathcal{B}$.

Each transition has a label of the form g/a, where $g \in \text{Bool}(\mathcal{I})$ must be true for the transition to be taken (g is the guard of the transition), and where $a \in \mathcal{O}^*$ is a conjunction of outputs that are emitted when the transition is taken (a is the action of the transition). State q is the source of the transition (q, g, a, q'), and state q' is the destination. A transition (q, g, a, q') will be graphically represented by $(q \xrightarrow{g,a} q')$.

The composition operator of two LTS put in parallel is the synchronous product, noted $\|$, and a characteristic feature of the synchronous languages. The synchronous product is commutative and associative. Formally $\langle \mathcal{Q}_1, q_{0,1}, \mathcal{I}_1, \mathcal{O}_1, \mathcal{T}_1 \rangle \| \langle \mathcal{Q}_2, q_{0,2}, \mathcal{I}_2, \mathcal{O}_2, \mathcal{T}_2 \rangle = \langle \mathcal{Q}_1 \times \mathcal{Q}_2, (q_{0,1}, q_{0,2}), \mathcal{I}_1 \cup \mathcal{I}_2, \mathcal{O}_1 \cup \mathcal{O}_2, \mathcal{T} \rangle$ with $\mathcal{T} = \left\{ \left((q_1, q_2) \xrightarrow[\wedge]{(g_1} g_2 \right)/(a_1 \wedge a_2)(q_1', q_2') \right) | (q_1 \xrightarrow[/]{g_1} a_1 q_1') \in \mathcal{T}_1, (q_2 \xrightarrow[/]{g_2} a_2 q_2') \in \mathcal{T}_2 \}$. Note that this synchronous composition is the simplified one presented in [37], and supposes that g and a do not share any variable, which would be permitted in synchronous languages like Esterel.

Here (q_1, q_2) is called a *macro-state*, where q_1 and q_2 are its two *component states*. A macro-state containing one component state for every LTS synchronously composed in a system S is called a *configuration* of S.

3.4 Discrete Controller Synthesis (DCS) on LTS

A system S is specified as a LTS, more precisely as the result of the synchronous composition of several LTS. $\mathcal{F}$ is the objective that the controlled system must fulfill, and $\mathcal{H}$ is the behavior hypothesis on the inputs of S. The controller C obtained with DCS achieves this objective by restraining the transitions of S, that is, by disabling those that would jeopardize the objective $\mathcal{F}$, considering hypothesis $\mathcal{H}$.

---

[2]http://bzr.inria.fr/pub/bzr-manual.pdf.

Both $\mathcal{F}$ and $\mathcal{H}$ are expressed as boolean equations. The set $\mathcal{I}$ of inputs of S is partitioned into two subsets: the set $\mathcal{I}_C$ of controllable variables and the set $\mathcal{I}_U$ of uncontrollable inputs. Formally, $\mathcal{I} = \mathcal{I}_C \cup \mathcal{I}_U$ and $\mathcal{I}_C \cap \mathcal{I}_U = \emptyset$. As a consequence, a transition guard $g \in \text{Bool}(\mathcal{I}_C \cup \mathcal{I}_U)$ can be seen as a function from $2^{\mathcal{I}_C} \times 2^{\mathcal{I}_U}$ into $\mathcal{B}$. A transition is controllable *if and only if* (iff) there exists at least one valuation of the controllable variables such that the boolean expression of its guard is false; otherwise it is uncontrollable. Formally, a transition $(q, g, a, q') \in \mathcal{T}$ is controllable iff $\exists X \in 2^{\mathcal{I}_C}$ such that $\forall Y \in 2^{\mathcal{I}_U}$, we have $g(X, Y) = \textit{false}$. In the proposed framework, the following function $S_c = \textit{make\_invariant}(S, E)$ from SIGALI is used to synthesize (i.e. *compute by inference*) the controlled system $S_c = S \| C$ where E is any subset of states of S, possibly specified itself as a predicate on states (or *control objective*) $\mathcal{F}$ and predicate on inputs (or *hypothesis*) $\mathcal{H}$. The function *make_invariant* synthesizes and returns a controllable system S_c, if it exists, such that the controllable transitions leading to states $q_i \notin E$ are inhibited, as well as those leading to states from where a sequence of uncontrollable transitions can lead to such states $q_i \notin E$. If DCS fails, it means that a controller of S does not exist for objective $\mathcal{F}$ and hypothesis $\mathcal{H}$. In this context, the present proposition relies on the use of DCS to synthesize a controller C, which makes invariant a safe set of states E in a LTS-based system where E is inferred by boolean equations defining a control objective and an hypothesis on the inputs. The controller C given by DCS is said to be *maximally permissive*, meaning that it doesn't set values of controllable variables that can be either true or false while still compliant with the control objective. Actually, the BZR compiler defaults these variables to true. Optimization can be done at this level if this type of decision is too arbitrary [11], but it goes beyond the scope of this work, which focuses on security, so the standard decision behavior given by BZR is kept. A smart home system, following the aforementioned execution principle, can now be designed using this framework.

4   Smart Home Model

From the various smart home presentations found in the related work, a smart home system for people with disabilities can be abstracted as a hierarchy of hardware and software components (dynamic or not), sensors, and effectors distributed among several interconnected rooms, helping a person with impairments to perform ADL. Showing how to specify all these features within a synchronous model is the aim of this section.

4.1   Dynamic Components

The top component of the hierarchy is the system itself. In accordance to the synchronous execution model, let S be the LTS of the system, taking inputs $\mathcal{I}$ from

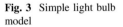

Fig. 3 Simple light bulb model

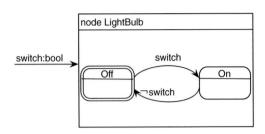

Fig. 4 Observable light bulb

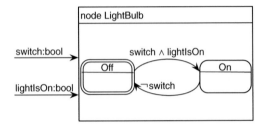

its effectors (buttons, touchscreens, controllable interfaces, etc.) and producing outputs $\mathcal{O}$ from its sensors (low level sensors, AI, any device producing notifications, etc.) each time it is triggered.

The smart home system is usually built upon several components, which can in turn be defined as LTS or LTS compositions if they are dynamic (i.e. they have multiple exclusive running modes) or as a set of synchronous equations if they have only one execution mode. Some components may be redundant and should not be specified more than once. For this case, BZR provides a *node* construct, in which LTS and synchronous equations can be defined to be instantiated. Figure 3 shows the graphical representation of such a node for a light bulb behavior definition.

Representing or not a component must be decided upon the following principle: if a component is concerned by a security rule, or if it can directly or indirectly influence a component concerned by a security rule, its behavior must be defined in the synchronous model. Moreover, if a behavior is modeled, it must also be observable. Regarding the example of the light bulb from Fig. 3, if its corresponding switch is set to ON or OFF,[3] then the bulb is supposed to respectively light up or shut down. This abstraction can work for a system with a relatively short life and built with new light bulbs. However, in the context of smart homes, a light bulb may fail at some point. In this model, the light bulb failure is not observable, so it does not correspond to reality. Being able to observe such a failure requires another component, like an appropriate sensor represented in Fig. 4 by the boolean variable *lightIsOn*. To keep track of the failure, it can be represented as an execution mode, cf. Fig. 5.

---

[3]The state of the switch is itself supposed to be known by the system.

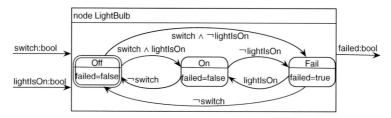

Fig. 5 Light bulb failure model

4.2 Person

Any person interaction with the system can be observed through its various types of sensors. However, in the very specific context of smart homes for disabled people, some characteristics of the person's behaviors and impairments can also influence security rules, and thus, have to be both observable and represented in the synchronous model.

As shown in the related literature, usual observable properties about a person are its position, mood, ADL and impairments. Position can be trivially defined as a LTS depending on how rooms are interconnected in the house. Let's suppose there are three rooms: a kitchen, connected to a bathroom and a bedroom. If sensors can determine the current position of a person, then the position evolution over time can be modeled by the LTS shown in Fig. 6. Observable behaviors in a smart home

Fig. 6 Position model

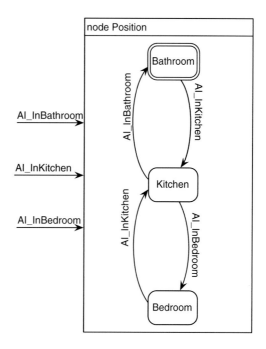

system are usually defined as a set of scenarios containing multiple steps and conditions to go from a step to another, so as to be processed by AI—in combination of events coming from the system—which can infer which step of which scenario the person is currently doing. Such a representation for scenarios makes them easy to be defined as LTS. And because a scenario can be aborted at any time by the person, modeling a scenario can follow the same principles presented for the observable failure of the light bulb. Figure 7 shows a LTS example representing the act of making coffee, evolving from step to step using AI notifications. Finally, mood and impairments are usually represented by boolean or numerical attributes, so they can be represented using synchronous equations. Evaluation of impairments for example, can come from various assessments such as the Global Deterioration Scale for Assessment of Primary Degenerative Dementia (GDSAPD) [38] which allocates a number between 1 and 7 depending on the cognitive decline (7 being very severe). We could also add additional disabilities such as "blind" or "deaf" which can be associated to booleans, cf. Fig. 8. It should be noted that this impairment model cannot evolve as it does not take inputs to influence the person's

Fig. 7 ADL "Make Coffee"

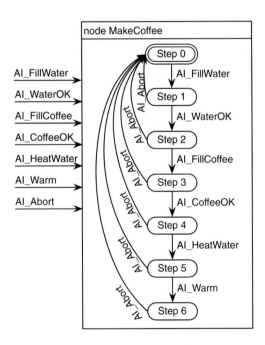

Fig. 8 Impairment model

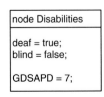

profile, so in the case this person is diagnosed with additional problems, this model should be changed accordingly, and recompiled. But this evolution could be represented with LTS and inputs as usual.

Using all these specifications, a smart home model can be completed by specific properties required by DSC, namely: designation of controllability within the model, and security constraints definition.

5 Applying DCS

When the various components and properties of a system are defined as behavior models (LTS, etc.) and synchronous equations, setting both the controllability and execution constraints enables the use of DCS.

5.1 Controllability

Controllability occurs naturally in the smart home domain. In the synchronous model, inputs are received each time the system is triggered, and these can come from both the environment—uncontrollable inputs $\mathcal{I}_U$ (e.g. a button is pressed by a human)—and the system itself—controllable inputs $\mathcal{I}_C$ (e.g. a device is forced to shut down by control system which is part of the execution loop).

For example, let's take a system allowing a third party application to control two failure-prone light bulbs so that they can be forced to light up or remaining lit even if their switch is turned off by a human. Figure 9 represents the designed by constraint controller of this small system, instantiating two times the LightBulb node (modified compared to Fig. 5 with a boolean variable c representing the aforementioned controllability), which takes amongst others the switches values as uncontrollable inputs $switch1, switch2 \in \mathcal{I}_U$ and the values given by the third party application as controllable boolean inputs $c1, c2 \in \mathcal{I}_C$. The statement $with$, declaring controllable variables, is actually implemented in BZR, which also allows to declare security constraints so that these variables can be valuated accordingly at each instant of the synchronous execution.

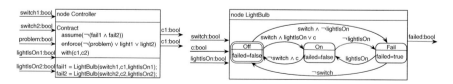

Fig. 9 Controllable light bulb model

5.2   *Constraints*

We consider two types of security constraints expressed as boolean synchronous expressions: (1) *Hypothesis*, which are supposed to remain true for all executions, and (2) *Guarantee*, which are enforced to remain true using controllable variables if and only if the *Hypothesis* stays true form the beginning of the execution.

For example, let's say we want to be sure that, for all possible executions, at least one light bulb is lit up if a problem (uncontrollable information coming from observation) arises: this can be specified using the guarantee ¬*problem* ∨ *light*1 ∨ *light*2 (cf. *enforce* statement). However, the system is not controllable with this rule alone: light bulbs can be in fail mode at the same time while the system receives a *problem*, and thus the guarantee cannot be fulfilled for this specific execution. This situation would be found automatically when applying DCS, which would fail to build a controller.

Now, let's say that the light bulbs can still fail but are supposed to be repaired quickly enough so that they don't fail at the same time. This is an example of fault tolerance: ultimately everything can fail but if there is enough redundancy we can safely state that not everything will fail at the same time. The hypothesis ¬(*fail*1 ∧ *fail*2) (cf. *assume* statement) represents this assumption in a synchronous boolean expression. Applying DCS using the BZR toolset on such a model gives back the C code of a controller taking $\mathcal{I}_U$ as inputs and providing the computation of $\mathcal{I}_C$ as outputs so that the system can now be executed, receiving both $\mathcal{I}_U$ and $\mathcal{I}_C$. DCS is able in this example to find automatically the correct controller code so that $c1$ and $c2$ can be valuated to *true* or *false* exactly when they should (e.g. when a problem arises, and lights are off, and *light*1 has failed, then $c2$ will be forced to false, etc.). From such a minimal example, we understand how DCS becomes interesting when the system's complexity in-creases while having to maintain its safety. If we add other failure-prone devices, impairment models, security constraints, etc. both designing and verifying the maximally permissive controller quickly start to be hard without appropriate tools.

6   Experiment

This section shows the application of DCS on the model of a smart home system to address various errors coming from the user's behavior or the system itself (failure of its components). The examples are built incrementally, i.e. they can be merged together into a model of a system on which DCS can be applied to synthesize a controller guaranteeing all user/component safety properties. They show four types of control behavior: (1) adaptation and (2) usage limitation to anticipate a user problem (known disabilities and potential behavior errors), and (3) adaptation and (4) usage limitation to anticipate components related problems (hardware failure). These types of control behavior are an answer to the smart home fault tolerance problems identified in [9].

6.1 Base Model

Before describing the aforementioned scenarios, let's represent the base model of a smart home system (cf Fig. 10). This model contains the elements concerned (directly or indirectly) by security constraints. Their behavioral definitions are given as a set of automata and synchronous equations following the BZR concepts. The model also specifies inputs and outputs, and follows the synchronous execution definition: each execution step consumes all inputs and computes all outputs through the specified equations and automata, the actual organization of computations inside a step being solved by the synchronous compiler. This specific view of the smart home system (i.e. its control related information) con-stitutes the designed-by-constraint definition of the controller that we will try to synthesize.

Point ① of this model represents the main node—the *controller node*—centralizing all incoming events and all necessary outputs. The first four inputs (*fail* and *repair*) are related to the failure and repair events of specific elements named islands, coming from a previous implementation of fault tolerance [17]. Islands are independent system monitoring several sensors and effectors. Here we consider two of them, instantiating (cf. Point ③) the generic *node island* definition given in point ⑦. They manage respectively (1) an iPad, a speaker, and (2) a light bulb (cf. point ③).

Beyond island events, the controller node also receives a notification when smoke is detected (*smoke*), when the kitchen room is in the dark (*dark*), when the radio is activated (*pushBtnRadio*), when the range hood fan in the kitchen is activated (*pushBtnRHF*), when the person receives a telephone call (*telephoneCall*), and when the stove burners 1 and 2 of the kitchen stove are set to then ON position (*activateSB1, activateSB2*).

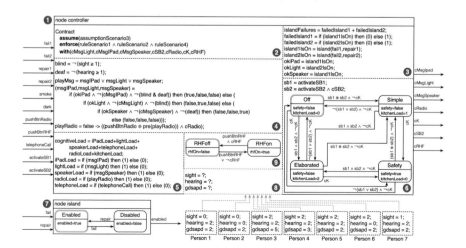

Fig. 10 Graphical representation of the defined by constraint controller

Point ⑥ shows a simplified behavior for the kitchen stove, allowing the activation of the two aforementioned stove burners, and being able to go in a safety mode if some problem is detected. When this mode is activated, the stove burners are set to OFF and the controller is notified (through the variable named *safety*) for actions to be derived.

Point ⑧ represents an abstract and simplified impairment model, which will be filled with a person actual data; here we will see what happens with seven persons given their information about their *sight* and *hearing* (integers from 0 to 2, meaning respectively "completely impaired" to "normal"), and also on their *GDSAPD* evaluation (integer from 1 to 7, meaning respectively "normal" to "severely mentally disabled"). These information will influence how the system should communicate—choosing between the iPad, the speaker, or the light—and when to communicate or not—*playMsg* being true or false—cf. point ④.

This communication with the user will also take care of the cognitive load: communicating with the person using the iPad, etc. increments the load by one unit, cf. point ⑤; if the kitchen stove is in use, this will also increase the load by one or two depending on the number of activated stove burners (cf. *kitchenLoad* from point ⑥); and finally, receiving a telephone call and listening to the radio are also considered as a cognitive loads. Thus, *telephoneLoad* and *radioLoad* (cf. point ⑤) increments the load by one depending on their activation (respectively influenced by the *telephone* input and the *playRadio* equation from point ④).

Finally, automata from ⑨ and ⑩ represent the activation behavior of the range hood fan and the kitchen light.

Now the model just needs a contract (which will be detailed in the next parts), so that DCS can be applied to eventually obtain an executable controller. This contract is given in point ② where, first, we make assumptions about the uncontrollability (*assumptionScenario3*), i.e. we define synchronous equations representing some key parts of the environment's behavior—this helps DCS to eliminate the verification of events combinations and sequences that are not supposed to happen—. And second, we specify rules that must remain true for all possible executions (*ruleScenario1,2,4*), with the help of seven controllable variables representing here the actual controllability of the system (i.e. they represent the real interface proposed by the smart home so that it can be influenced through the use of a computing system). These controllable variables are valuated internally by the synthesized controller (obtained through DCS) and provided as outputs, so that the system can react at each control step. When these variables are forced to false, *cMsg(IPad/Light/Speaker)* indicate which prompting system has to be used (respectively the iPad, the light, and the speaker), *cRadio* can force the radio to turn off, *cK* can set the kitchen stove in a safety mode, *cSB2* can prevent the use of the second stove burner and *cRHF* can activate the range hood fan.

We will now incrementally set four types of safety rules to tolerate errors or to adapt upon dangers, and detail their effects when the smart home system under control is used by people with different disabilities and different levels of impairments.

6.2 Scenario 1: User Assistance (Adapted Prompting)

Description and objective This first scenario aims at showing the added value of DCS when designing a controller to adapt the way the smart home communicates with the user. Guaranteeing the safety of the controller (with respect to rules) is made through the verification aspect of DCS, just like in classical formal verification algorithms (e.g. Model Checking); but knowing if such a controller actually exists is addressed by the very specific aspect of DCS: synthesis from constraints. Let's specify a first constraint, for example we want to be sure that when smoke is detected in the kitchen (*smoke* variable is *true*), then the range hood fan activates and a prompt information is provided through an adapted media (iPad, speaker or light-bulb). To reflect this in the model, we set the following equation for *ruleScenario1*:

$$\text{ruleScenario1} = \neg(\text{smoke}) \lor (\text{rhfon} \land \text{playMsg}) \tag{1}$$

We now take the cases of persons 1 and 2: will the smart home be controllable, i.e. will it be able to cope with this constraint for all possible execution? Let's apply DCS with each user profile on this specification—containing the base model, the user profile, and this first rule (the other ones being set to *true* for the moment)—and let's simulate a scenario where the user activates the two stove burners to start cooking some food, but smoke gets detected when the food is starting to burn.

Comments on scenario execution DCS fails when applied with the profile of person 2. This person is actually both blind and deaf, so there is no appropriate communication media in case of a problem (e.g. when smoke is detected). Technically: variables *msg(IPad/Speaker/Light)* can never be set to *true* what-ever the values given to the controllable variables *cMsg(IPad/Speaker/Light)* because *blind* and *deaf* are *true*; thus *playMsg* can never be *true*. This leads to the fact that *ruleScenario1* can be *false* if *smoke*—an uncontrollable variable given as input—is *true*, which is of course not permitted (*ruleScenario1* has to be enforced to *true* for all executions). Because of this, the smart home system is not controllable for this first rule and this is the reason why DCS fails, meaning that the system has to be reworked (e.g. by adding adapted medias), before person 2 can actually use it safely.

However, DCS succeeds when applied with the profile of person 1, meaning that a controller has been found so that *ruleScenario1* will always remain *true* for all possible execution. Figure 11 shows the simulation results, highlighting important events. It has to be noted that only the relevant events and steps are represented:

Fig. 11 Execution of scenario 1 for person 1

		Person 1					
		Step 1	Step 2	Step 3	Step 4	Step 5	Step 6
Inputs	smoke	0	0	1	1	0	0
	activateSB1	1	1	1	1	1	1
	activateSB2	0	1	1	1	1	1
	pushBtnRHF	0	0	0	1	0	1
Out.	cMsgSpeaker	1	1	0	0	1	1
	cRHF	1	1	0	0	1	1

missing inputs are *false*, missing outputs are *true*, and a new step is represented only when something changes from the previous one (new inputs/outputs or internal state modification).

- Steps 1 and 2: Person 1 activates respectively the stove burners 1 and 2 to start cooking.
- Step 3: Some smoke gets detected by a sensor that valuate the *smoke* input to *true*; the controller reacts by setting the controllable variables *cMsgSpeaker*, *cK* and *cRHF* to *false*, which has the following effects:

  - the range hood fan activates
  - a message about the smoke is played using the speaker; physically, the system is able to select the appropriate message and media, knowing that *smoke* is true and *cMsgSpeaker* is forced to *false*.

  This way, *ruleScenario1* remains true.

- Step 4: The person tries to deactivate the range hood fan, however, because smoke is still detected, the fan has to stay activate; deactivation is actually prevented by forcing cRHF to remain false in this step, avoiding to take the transition from step *RHFon* to *RHFoff*.
- Step 5: No smoke is detected anymore, and no controllable variable has to be forced to *false*; this has the following effect: the message about smoke is stopped being played.
- Step 6: the person tries to deactivate the range hood fan, which is this time permitted because *cRHF* is not forced to *false*.

This scenario has shown the interest of using DCS in this context to both verify that the smart home is able to adapt to a person's disabilities and generate a controller to manage the smart home adaptation behavior.

6.3 Scenario 2: User Error Prevention (Simultaneous Devices Usage Limitation)

Description and objective This second scenario shows the advantage of using DCS to put limitations on how the various devices in a smart home can be used, depending on the user's profile. As an example of such a case, we will focus here on the compound cognitive load, due to simultaneous device usage by a person, and show how the smart home system gets configured to prevent a cognitive overload. Let's specify a second constraint, such that the cognitive load cannot exceed 2 and 3 units when the person's GDSAPD is respectively evaluated to 5 and 3. Regarding the adaptation possibilities, the radio can be turned o automatically and the kitchen stove cannot be used on *Elaborated* mode (only the stove burner 1 can be activated at most) to reduce the cognitive load when it is necessary. We set this as the following equation for *ruleScenario2*:

$$\text{ruleScenario2} = (\neg(\text{gsdapd} \geq 5) \vee (\text{cognitiveLoad} \leq 2)) \wedge$$
$$(\neg(\text{gdsapd} \geq 3) \vee (\text{cognitiveLoad} \leq 3)) \tag{2}$$

Like in the example given in the first scenario, if the smart home cannot be adapted to a person (due to disabilities) for all possible executions, DCS will fail to find a controller. So we will take the cases of persons 3, 4 and 5—for which DCS succeeds—and simulate a scenario where the person is listening to the radio, then activates the stove burners 1 and 2 to start cooking, but then receives a telephone call (Fig. 12).

Comments on scenario execution Fig. 12 shows the simulation results, where we can see how the smart home gets adapted to cope with the differences in abilities to deal with cognitive load between the three persons. Person 5, having a GDSAPD lower than 3, should be able to deal with a high cognitive load, but it is not the case for persons 4 and 3, and this is why they experience some limitation when using some devices simultaneously in this controlled smart home.

- Step 1: Persons 3, 4 and 5 start listening to the radio; cognitive load is then set to 1 unit, which is correct for everyone.
- Step 2: Persons 3, 4 and 5 activate the first stove burner; this increments the cognitive load by 1 unit, which is now the acceptable limit for person 3, having a GDSAPD equal to 5 units.
- Step 3: Person 4 and 5 activate the second stove burner, now the cognitive load is set to 3 units, the acceptable limit for person 4 (having a GDSAPD equal to 3 units); but when person 2 tries to activate the second stove burner by setting the *activateSB1* switch to "on" (*true*), the radio goes o because the controller forces *cRadio* to be *false*, thus keeping the cognitive load below the acceptable limit, as required by *ruleScenario2*. It is interesting to note here that preventing the second stove burner to start by forcing *cSB2* to be *false* would also have been a correct response from the controller. The order in which the controllable variables are set actually depends on the order they are declared in the BZR program. Here *cSB2* is declared before *cRadio*, so when a value is asked for *cSB2* in a step, the value of *cRadio* is not decided yet, and in this example *cSB2* has no reason to be forced to false, because there is still a solution to comply with the rules (i.e. by setting *cRadio* to false). Inverting the declarations of *cSB2* and *cRadio* would have let the radio "on" and prevented the use of the second kitchen stove for person 3 in this step.

		Person 3				Person 4				Person 5			
		Step 1	Step 2	Step 3	Step 4	Step 1	Step 2	Step 3	Step 4	Step 1	Step 2	Step 3	Step 4
Inputs	pushBtnRadio	1	0	0	0	1	0	0	0	1	0	0	0
	telephoneCall	0	0	0	1	0	0	0	1	0	0	0	1
	activateSB1	0	1	1	1	0	1	1	1	0	1	1	1
	activateSB2	0	0	1	1	0	0	1	1	0	0	1	1
Out.	cRadio	1	1	0	0	1	1	1	0	1	1	1	1
	cSB2	1	1	1	0	1	1	1	1	1	1	1	1

Fig. 12 Execution of scenario 2 for persons 3, 4 and 5

- Step 4: a telephone call is received in this step, which increments the cognitive load by one, and this cannot be prevented (there is no controllability on receiving telephone calls for this smart home, at least in the model we have defined). Person 5 can receive the call whilst continuing to cook and to listen to the radio. However letting the cognitive load going to 4 units is not permitted for person 4, so the smart home has to react to not let this happen: this is why the radio gets deactivated (and not the stove burner 2 for the same reason explained in step 3 for person 3). Finally, this telephone call impacts the smart home usability for person 3 to keep the cognitive load to an acceptable level: the controller forces *cSB2* to be false here, thus deactivating the second stove burner and keeping the cognitive load to 2 units.

This scenario has shown an example on how the person's static profile can be used to prevent user errors (here by keeping the cognitive load below an adapted level) by configuring the smart home with the help of DCS which provided a corresponding smart home controller.

6.4   Scenario 3: Component Failure (Redundancy)

Description and objective This third scenario shows the application of DCS to solve a problem that could arise from a previous implementation of fault tolerance in our smart home system: in our architecture presented in [17], we "did install industrial grade material [...] to avoid hazardous situations [for example where] the resident cannot turn on the light due to a system failure."; thus we connected our various sensors and effectors to four independent fault-tolerant islands so that "if a block falls, only the [connected equipments] will be affected". Sensors and effectors are critical safety elements, so if their connected island can fail, do we have enough redundancy? To generalize, having enough redundancy in a given system means that, for all possible execution of this system and in case of a failure, there is always a solution to keep it running correctly; it means here that the smart home system remains adapted to the person's impairments. In order to keep the base model small and visually clear, we consider a simplification of our own redundancy implementation where only islands can fail but not the other devices (lights, sensors, etc.). Island failure is a kind of uncontrollable event (the system can do nothing to prevent this), so it cannot be represented as a control rule. However we will assume here that the case where two islands are disabled in the same step should never happen, but we still want to tolerate one and only one failure at most. This hypothesis can be represented in the assume part of the contract, by giving the following equation to *assumptionRule3*:

$$\text{assumptionRule3} = \text{islandfailures} \leq 1 \qquad (3)$$

In this scenario, we will see that having an island failure may have different impacts on the smart home behavior regarding who is actually living in. It starts when the island number 1 fails. Then the user activates the two stove burners to start cooking, but smoke gets detected which triggers a message (using an appropriate media) and the user deactivate the stove burners. At some point, no more smoke is detected and the island number 1 gets repaired. Then the user re-starts cooking, but smoke gets detected again and an new message has to be communicated. We take persons 1 and 6 for this scenario and apply DCS (Fig. 13).

Comments on scenario execution As shown in Fig. 13 for person 1, DCS does not find an appropriate controller. Indeed, if island number 1 fails, then the speakers cannot be activated (cf. Point ③, *island1IsOn* being *false* means *okSpeaker* is *false* too, and then *msgSpeaker* cannot be *true*); however this is the only acceptable communication media when a person is blind (but not deaf) which is the case of person 1 (cf. Point ④, if *blind* is *true*, then *msgLight* and *msgIPad* cannot be *true*); so if *msg(IPad/Speaker/Light)* are all false, then *playMsg* becomes false and this can be problematic: if a message has to be communicated to the user because of a problem, such as a smoke problem as specified in ruleScenario1, there is no media available and because this situation cannot be prevented by any available controllability then DCS fails, meaning that the redundancy implementation has to be reworked.

For person 6 however, DCS succeeds and the steps results of the aforementioned scenario can be seen in Fig. 13. It has to be noted that if both islands fail at the same time, then no media can be selected anymore, and in case of a smoke problem this would violate *ruleScenario1*. But DCS ignores this case, as we have defined that a double island failure should not happen at the same time in the assume part of the contract.

– Step 1: The scenario starts when island number 1 fails; this disables the use of the iPad and the speaker as communication devices.
– Steps 2 and 3: Person 6 activates the two stove burners to start cooking.
– Step 4: Smoke gets detected from the environment (this enables the range hood fan but it is not relevant in this scenario), and a message has to be communicated in order to keep *ruleScenario1* to true. Person 6 being deaf, the iPad cannot be used as it requires to have correct sight and hearing (cf. Point ④); so only the

| | | Person 6 | | | | | | | |
		Step 1	Step 2	Step 3	Step 4	Step 5	Step 6	Step 7	Step 8
Inputs	fail1	1	0	0	0	0	0	0	0
	repair1	0	0	0	0	0	0	1	0
	smoke	0	0	0	1	1	0	0	1
	activateSB1	0	1	1	1	0	0	1	1
	activateSB2	0	0	1	1	0	0	1	1
Outputs	cMsgLight	1	1	1	0	0	1	1	1
	cMsgIPad	1	1	1	1	1	1	1	0
	cMsgSpeaker	1	1	1	1	1	1	1	1

Fig. 13 Execution of scenario 3 for person 6

light remains, and this is why *cMsgLight* is forced to *false* by the controller, this way a message about the smoke problem can be communicated through the light.

- Step 5: Person 6 sets the two stove burners' switches to "off".
- Step 6: No smoke is detected, so the controller does not continue to keep *cMsgLight* to *false* (having *smoke* to *false* keeps *ruleScenario1* to *true*) thus the message about smoke can be stopped.
- Step 7: Island 1 has been repaired and the two stove burners are switched "on" again.
- Step 8: Smoke is detected anew: This time, the message about smoke gets communicated through the iPad, because (1) island number 1 is operational and (2) the controllable variable *cMsgIpad* is declared after *cMsgLight*, explaining why *cMsgLight* remains true (not forced by the controller), because *cMsgIpad* can always be forced to false after (which is the case in this step), thus keeping *ruleScenario1* to true.

Usage of DSC in this scenario is especially powerful: instead of trying to define redundancy generically for all types of users, we can reduce costs by defining more or less redundant component combinations for different users and applying DCS to guarantee that a specific redundant installation is safe for a specific user profile. Moreover, adjusting hypothesis on the components' quality is simply a question of setting a boolean equation in the assume part. For example, if we have five islands monitoring our components, and want to tolerate a maximum of three failures at the same time and see if our system is still stable, we just have to set (*islandfailures* ≤ 3) in the *assume* part and apply DCS to know if the smart home is actually stable (controllable) in this context and get the associated controller.

6.5   Scenario 4: Component Adaptation (Degraded Mode)

Description and objective An alternative solution to cope with hardware failures, besides using redundancy, is to modify the way the remaining operational components can be used. If we take the islands failures example, and want to cope with a double failure—which creates a communication problem when smoke is detected -, then we can program a controller to degrade what can create smoke (i.e. the kitchen stove by setting *cK* to *false*), and either remove *ruleScenario1* or assume that *smoke* remains *false* when *Safety* mode is active. But it would mean that the kitchen stove would be forced to remain in *Safety* mode for all possible executions just because smoke could happen, which is not acceptable. Instead, we want to find an example showing how DCS can be useful to build a controller helping to anticipate a hardware failure by modifying the remaining active components in the case where no redundancy is available. Let's say we only have one lightbulb in the kitchen. At night, if the lightbulb fails—and whatever its controllability (i.e. the lightbulb can be switched "on" or "off" automatically)—then the kitchen goes completely dark. Now let's build an new safety rule such that, if the user has limited sight, the kitchen stove cannot be used during the night when the lightbulb is not activated. Of course, because the light bulb has no redundancy, we cannot create a rule such that when the kitchen is used during the night, then the light should go "on" if sight is limited: this model would not comply with the reality if we consider that the lightbulb can fail (and thus cannot go "on" at some undetermined moment). So instead, we use a light sensor providing the value of a variable named *dark*— indicating if there is enough light in the room (*false*) or not (*true*)—and this value explicitly impacts the way the kitchen can be used through the following synchronous equation attributed to *ruleScenario4*:

$$\text{ruleScenario4} = \neg(\text{sight} == 1) \vee (\neg\text{dark}) \vee (\text{kitchenLoad} == 0) \quad (4)$$

This means that if the user's sight is evaluated to 1 (visually impaired but not blind), and if the kitchen is in the dark, then the smart home has to adapt itself so that the kitchen stove cannot be used (it is either in *Off* or *Safety* mode, the only modes where *kitchenLoad* equals 0).

We now define a scenario where we start at night, the kitchen is in the dark, and the user activates the stove burner 1; then the user presses the light button (this event is not given to the controller because no information about the actual lightbulb activation can be safely derived from it), but the lightbulb fails in the following instant.

Depending on the user's profile, different control results happen when this scenario is played, as shown is Fig. 5 for persons 7 and 5 (for which DCS succeeded). This example relates to Scenario 1 regarding the explicit constraint definition on environment events (smoke value directly impacts the activation of the range hood fan), to Scenario 2 regarding the dynamic adaptation to multiple users profiles, and to Scenario 3 regarding the hardware failure example (islands can fail, their failures impact the system's behavior) (Fig. 14).

Comments on scenario execution Following Fig. 14, person 7 being visually impaired, the smart home system adapts itself consequently. However, person 5 having a good sight, the smart home does not interfere with the kitchen stove usage, whatever the light condition.

		Person 6							
		Step 1	Step 2	Step 3	Step 4	Step 5	Step 6	Step 7	Step 8
Inputs	fail1	1	0	0	0	0	0	0	0
	repair1	0	0	0	0	0	0	1	0
	smoke	0	0	0	1	1	0	0	1
	activateSB1	0	1	1	1	0	0	1	1
	activateSB2	0	0	1	1	0	0	1	1
Outputs	cMsgLight	1	1	1	0	0	1	1	1
	cMsgIPad	1	1	1	1	1	1	1	0
	cMsgSpeaker	1	1	1	1	1	1	1	1

Fig. 14 Execution of scenario 4 for persons 7 and 5

- Step 1: Both users are in the kitchen, in the dark.
- Step 2: Both of them press a button to activate the stove burner 1, but the actual activation is prevented for person 7: the controller forces cK to *false*, which prevents the kitchen stove to go in *Simple* execution mode where *kitchenLoad* would be equal to 1 instead of 0, thus violating *ruleScenario4*.
- Step 3: They both activate the lightbulb, and the light sensor reacts by setting *dark* to *false*. This allows the actual activation of the stove burner 1 for person 7.
- Step 4: The kitchen returns in the dark, as indicated by the *dark* value (*true*), even if neither person 7 nor 5 actually touched the light switch to turn it off: the lightbulb has failed and the controller reacts on the *dark* value—instead of the actual light switch position—for person 7 by setting the cK controllable variable to *false*, thus placing the kitchen stove to its safety execution mode.

In the context of usage limitation to anticipate components related problems, this example shows again the advantage of being able to define the system's controllability by constraint, instead of giving its actual implementation: here the kitchen stove is actually controllable in the sense that an internal system (the controller) can act on it to prevent the activation of its stove burners, but the kitchen model does not have to define explicitly under which conditions it has to react; the actual control implementation (valuation of the controllable variables) is obtained by synthesis given the global constraints defined by the programmer in the model of the system's components. States or behaviors of one or several components (e.g. the light and impairment profile) can have an impact on the controllability of other components (e.g. the kitchen stove) without requiring to define explicitly this controllability.

6.6 Evaluation

Our approach is compared in Fig. 15 to the most relevant ones already discussed in the "related work" section of this document. Beforehand, we had implemented some redundancy mechanisms in our own smart home test lab [17], so that not all sensors/effectors would be controlled by a single machine (an Island) because it would have consisted in a single point of failure. However, we could not be exactly sure that for every susceptible failure, the use of redundant elements (islands) would

	Smart home context	Smart home modeling	Fault tolerance	Disability context	Adaptation to a person	Methodology, concrete scenarios and examples	Verification	Synthesis
Dumitrescu et al. 2010			x				x	x
Bouchard et al. 2012	x		x	x	x	x		
Chetan et al. 2005	x		x	x				
Corno et al. 2013	x	x					x	
Lapointe et al. 2012					x	x		
Zaho et al. 2012	x	x					x	x
Guillet et al. 2013	x	x	x	x	x		x	x
This study	x	x	x	x	x	x	x	x

Fig. 15 Comparison to other approaches

be sufficient to keep the smart home safe for a particular person (as several sensors and/or effectors not managed by the new island would have been disabled). The way we connect our sensors and effectors to multiple islands could actually be safe for a person in case of a failure but not for another one with different disabilities, and this could be hard to find and verify without appropriate tools able to solve this combinatorial problem. Redundancy without verification was indeed not sufficient.

As we were trying to solve this failure problem, we compared our approach to closely related ones regarding smart homes for disabled people, and learned—especially from [9]—that failures could be of multiple types in this context and redundancy itself was not the only solution to address them. This is why we became interested in the more general problem of fault tolerance for this kind of smart homes, and we would base our use cases on their studies to show how the different types of failures could be addressed. But unlike these approaches, we would complete ours by verifying it.

Designing smart home models models such that they can be verified has been done several times. One of the most related approaches regarding modeling and verification is [29], where a smart home is modeled using a formal representation (State Charts), and safety properties defined so that formal verification tools can be employed to guarantee that these properties are ensured for all possible execution. However, we already discussed the problem of modeling the entire system to apply verification. This why we kept the formal modeling approach, but took an approach based on synthesis to address this combinatorial problem by solving it automatically from constraints, instead of trying to find (and verify) a complete solution manually. Our expertise with synthesis techniques comes from previous work in the domain of reconfigurable hardware architectures, where DCS was proven useful to build formal reconfiguration controllers. Still in the hardware context, DCS was also employed to carry out computations on failure-prone processors, giving us inspiration on how to actually use DCS to manage fault tolerant smart homes.

Usage of synthesis techniques in the smart home context is very recent. First results can be seen in [34, 35] which present the use of DCS respectively from a generic point of view (in the context of the Internet of Things) and from a specific one regarding fault tolerance. However, being preliminary, they both lack of concrete use cases, implementation and results. This proposal makes a contribution over them by giving these elements: it presents a detailed methodology to create smart home controllers by synthesis, using BZR for smart home elements modeling, and shows use cases with realistic scenarios that we tested in our lab to demonstrate the relevance and advantages of using DCS for addressing the various fault tolerance problems (identified in [9]) that may occur in a smart home dedicated to disabled people.

7   Conclusion and Perspectives

Safety and security services are essential requirements for many pervasive computing systems. This is especially true for smart homes dedicated to people with disabilities, where security constraints prevail. They represent a pervasive systems category where safety is actually a very critical property: the person living in such a house is usually frail and is not supposed to be able to cope with errors; implication of failures can range from user annoyance to hazardous situations.

Correct adaptation behavior—so that the smart home remains safe whatever the conditions of execution—is both difficult to design and verify. While verification has been addressed multiple times, uses of synthesis techniques in this context to guarantee a safe behavior (employing formal verification) while simplifying the design (which is derived from constraints) are still rarely encountered and lack of examples showing how they can be used to solve practical problems. In this context, this proposal makes a contribution by providing a design methodology, relying on DCS, and backed by scenarios examples, to build smart home controller systems guaranteeing safety properties.

The results validating the proposal present both modeling and executions parts for different scenarios. They especially focus on fault tolerance as a safety property, and show how to deal with four types of typical control needs in this context: adaptation and usage limitation for users problems and components failures. With these results, obtained by rigorous experiments (real scenarios, executed in our smart home test lab), we demonstrated that the synchronous paradigm (on which BZR is based) and DCS tools (such as Sigali) are a relevant to design and compute the controller of a smart home system, in the context of fault tolerance.

In the end, the proposed methodology allows us to solve a simple but crucial question: can this smart home be adapted to this person, for every failure situation that can be derived from its model? A negative answer implies that a safety constraint (defined in the model) can be violated, and this cannot be prevented: the smart home itself has to be modified (by adding more redundancy, removing dangerous elements or executions modes, etc.) because no correct adaptation controller exist. However, a positive answer to this question automatically gives back the code of a correct control system, to be connected (inputs and outputs) and executed within the corresponding smart home so that it can actually be adapted dynamically.

As a perspective, the current methodology could be improved by defining an adequate abstraction level so that smart home designers would not even have to learn about BZR. For example, such an abstraction has been implemented in the reconfigurable embedded systems domain (cf. [11]) to allow designers to specify reliable reconfiguration controllers using only a UML profile (high level abstraction); models built with this profile could be transformed into a synchronous representation based on BZR to make DCS applicable transparently, thus giving back the executable code of their specified-by-constraints controllers.

References

1. Ramos, C., Augusto, J.C., Shapiro, D.: Ambient intelligence: the next step for artificial intelligence. Intell. Syst. IEEE **23** (2008)
2. Carberry, S.: Techniques for Plan Recognition. User Model. User-Adap. Inter. **11**, 31–48 (2001)
3. Novak, M., Binas, M., Jakab, F.: Unobtrusive anomaly detection in presence of elderly in a smart-home environment. In: ELEKTRO (2012)
4. Bouchard, B., Giroux, S., Bouzouane, A.: A keyhole plan recognition model for Alzheimer's patients: first results. J. Appl. Artif. Intell. (AAI) **21**, 623–658 (2007)
5. Fortin-Simard, D., Bouchard, K., Gaboury, S., Bouchard, B., Bouzouane, A.: Accurate passive RFID localization system for smart homes. In: IEEE 3rd International Conference on Networked Embedded Systems for Every Application (NESEA), pp. 1–8 (2012)
6. Lapointe, J., Bouchard, B., Bouchard, J., Potvin, A., Bouzouane, A.: Smart homes for people with Alzheimer's disease: adapting prompting strategies to the patient's cognitive profile. In: Proceedings of the 5th International Conference on Pervasive Technologies Related to Assistive Environments, pp. 30:1–30:8. New York, NY, USA, ACM (2012)
7. Pigot, H., Mayers, A., Giroux, S.: The intelligent habitat and everyday life activity support. In: 5th International Conference on Simulations in Biomedicine, avril 2003. Slovenie (2003)
8. Bulow, J.: An economic theory of planned obsolescence. Q. J. Econ. **101**, 729–749 (1986)
9. Chetan, S., Ranganathan, A., Campbell, R.: Towards fault tolerance pervasive computing. Technol. Soc. Mag. **24** (2005)
10. Ramadge, P.J.G., Wonham, W.M.: The control of discrete event systems. Proc. IEEE **77**, 81–98 (1989)
11. Guillet, S., de Lamotte, F., Le Griguer, N., Rutten, É., Gogniat, G., Diguet, J.P.: Designing formal recon guration control using UML/MARTE. In: 7th International Workshop on Reconfigurable Communication-centric Systems-on-Chip (Re-CoSoC), pp. 1–8 (2012)
12. Pigot, H., Lefebvre, B., Meunier, J.G., Kerhervé, B., Mayers, A., Giroux, S.: The role of intelligent habitats in upholding elders in residence. In: 5th International Conference on Simulations in Biomedicine, pp. 497–506 (2003)
13. Latfi, F., Lefebvre, B., Descheneaux, C.: Ontology-based management of the tele-health smart home, dedicated to elderly in loss of cognitive autonomy. In: Workshop on OWL: Experiences and Directions, pp. 1–10 (2007)
14. Augusto, J.C., Nugent, C.D.: Smart homes can be smarter. In: In Designing Smart Homes— The Role of Artificial Intelligence (2006)
15. Patterson, D.J., Kautz, H.A., Fox, D., Liao, L.: Pervasive computing in the home and community. In Bardram, J.E., Mihailidis, A., Wan, D. (eds.) Pervasive Computing in Healthcare, pp. 79–103. CRC Press (2006)
16. Mihailidis, A., Boger, J., Canido, M., Hoey, J.: The use of an intelligent prompting system for people with dementia. Interactions **14** (2007)
17. Bouchard, K., Bouchard, B., Bouzouane, A.: Guidelines to efficient smart home design for rapid AI prototyping: a case study. In: Proceedings of the 5th International Conference on Pervasive Technologies Related to Assistive Environments, New York, NY, USA, ACM (2012)
18. Bouchard, K., Bouchard, B., Bouzouane, A.: Discovery of topological relations for spatial activity recognition. In: Proceedings of the IEEE Symposium Series on Computational Intelligence (SSCI 2013), pp. 1–8 (2013)
19. Benveniste, A., Caspi, P., Edwards, S.A., Halbwachs, N., Le Guernic, P., de Simone, R.: The synchronous languages 12 years later. Proc. IEEE **91**, 64–83 (2003)
20. Marchand, H., Bournai, P., Borgne, M.L., Guernic, P.L.: Synthesis of discrete-event controllers based on the signal environment. Discrete Event Dyn. Syst. **10**, 325–346 (2000)

21. Dumitrescu, E., Girault, A., Marchand, H., Rutten, É.: Multicriteria optimal dis-crete controller synthesis for fault-tolerant real-time tasks. In: Workshop on Dis-crete Event Systems, WODES'10, pp. 366–373. Berlin, Germany (2010)
22. Delaval, G., Marchand, H., Rutten, E.: Contracts for modular discrete controller synthesis. In: Proceedings of the ACM SIGPLAN/SIGBED 2010 Conference on Languages, compilers, and Tools for Embedded Systems, pp. 57–66. New York, NY, USA, ACM (2010)
23. Kilgore, C., Peitz, M., Schmid, K.: System Requirements Document for Safe Home. Research report, Iowa State University (2004)
24. Le Lann, G.: The Ariane 5 flight 501 failure—a case study in system engineer-ing for computing systems. Technical report, REFLECS—INRIA Rocquencourt (1996)
25. Jaygarl, H., Denner, A., Pham, N.: Software requirements and specification document for smart home notification and calendering system. Research report (smart home project, Iowa State University), pp. 1–45 (2008)
26. Picard, R.W.: A ective computing. Technical report (1995)
27. Schmidtke, H.R., Woo, W.: Towards ontology-based formal verification methods for context aware systems. In: Proceedings of the 7th International Conference on Pervasive Computing, pp. 309–326. Springer, Berlin, Heidelberg, (2009)
28. Corno, F., Sanaullah, M.: Formal verification of device state chart models. In: 7th International Conference Intelligent Environments (2011)
29. Corno, F., Sanaullah, M.: Modeling and formal verification of smart environments. Secur. Commun. Netw. (2013)
30. Kephart, J.O., Chess, D.M.: The vision of autonomic computing. IEEE Comput. (2003)
31. Cassandras, C.G., Lafortune, S.: Introduction to Discrete Event Systems. Springer, Berlin (2006)
32. Delaval, G., Rutten, E.: Reactive model-based control of reconfiguration in the fractal component-based model. CBSE (2010)
33. Bouhadiba, T., Sabah, Q., Delaval, G., Rutten, E.: Synchronous control of reconfiguration in fractal component-based systems—a case study. In: Proceedings of the International Conference on Embedded Software. EMSOFT (2011)
34. Zhao, M., Privat, G., Rutten, É., Alla, H.: Discrete control for the internet of things and smart environments. In: 8th International Workshop on Feedback Computing, In conjunction with ICAC (2013)
35. Guillet, S., Bouchard, B., Bouzouane, A.: Correct by construction security approach to design fault tolerant smart homes for disabled people. In: EUSPN (2013)
36. Marchand, H., Samaan, M.: Incremental design of a power transformer station controller using a controller synthesis methodology. IEEE Trans. Softw. Eng. 26(8), 729–741 (2000)
37. Altisen, K., Clodic, A., Maraninchi, F., Rutten, É.: Using controller-synthesis techniques to build property-enforcing layers. In: Proceedings of the 12th European conference on Programming, pp. 174–188. Springer, Berlin, Heidelberg (2003)
38. Reisberg, B., Ferris, S.H., Crook, T.: Signs, symptoms and course of age-associated cognitive decline. In Corkin, S., Davis, K.L., Growden, J.H. (eds.) Aging: Alzheimer's Disease: A Report of Progress, pp. 177–181. Raven Press (1982)

Smart Homes in the Era of Big Data

Kevin Bouchard, Sebastien Gaboury, Bruno Bouchard,
Abdenour Bouzouane and Sylvain Giroux

1   Introduction

The rapid evolution of sensing technology and the increasing power of computation
have resulted in the emergence of smart homes [1]. Those environments, which
implement the old dream of ubiquitous computing first described by Weiser [2] on a
small scale, are improved living spaces equipped with distributed sensors and
effectors hidden from the view of the residents. Smart home has been a very active
area of research through the last two decades, which resulted into various applications and philosophies of implementation and design. In particular, it has been
seen as a way to enhance the quality of life of residents by automating daily tasks
[3] and optimizing power consumption [4]. Another very important trend is the
assistance of residents in their daily life activities with the help of smart home
technology [5, 6]. Researchers envision a future were persons afflicted by a cognitive disease, such as mild dementia or head trauma, could pursue a
semi-autonomous life at their residence for an extended period. To achieve that goal
however, many challenges need to be addressed by the community. On the hardware side, researchers must be able to select the sensing technologies best adapted
to this context [7]. A part of the community relies on video cameras [8, 9] which are

K. Bouchard (✉) · S. Giroux
DOMUS, Universite de Sherbrooke, Sherbrooke, QC J1H 1H9, Canada
e-mail: Kevin.Bouchard@usherbrooke.ca

S. Giroux
e-mail: Sylvain.Giroux@usherbrooke.ca

S. Gaboury · B. Bouchard · A. Bouzouane
LIARA, Universite du Quebec a Chicoutimi, Chicoutimi, QC G7H 2B1, Canada
e-mail: Bruno.Bouchard@uqac.ca

A. Bouzouane
e-mail: Abdenour.Bouzouane@uqac.ca

© Springer International Publishing Switzerland 2016
K.K. Ravulakollu et al. (eds.), *Trends in Ambient Intelligent Systems*,
Studies in Computational Intelligence 633, DOI 10.1007/978-3-319-30184-6_5

power full sensors but raise concerns with the residents regarding their privacy [10]. The other part of the community reposes on ubiquitous sensors of various kinds (motion, ultrasound, electromagnetic contacts, etc.) [5, 11, 12]. The advantage is that these sensors can be hidden very well, but they also provide the system with fewer information from hybrid data. Another challenge of assistive smart home is the recognition of the activities of daily living (ADLs) performed by one or many residents. To that end, researchers first tried to develop algorithms based on mathematical logics such as the first order logic [13] or the possibility theory [14]. Other have relied on statistical models such as the Bayesian network [15] or Hidden Markov Model (HMM) [16]. More recently, due to the difficulty of constructing plans' library for those algorithms, researchers have worked toward the development of data mining techniques for activity recognition and learning [17, 18]. Finally, important challenges also present themselves to the researchers regarding the development of assistance algorithm. Many researchers have worked on the selection of the best prompting technology according to the profile of the person [19], but research on context awareness for the adaptation of the software have also been blossoming [20].

In the state of the art of smart home research, these challenges are well-known and have been documented for years [1, 21, 22]. However, computer science is now entering in the era of data due to the continuous trends of interconnection and ubiquity of systems. The so-called context of Big Data is creating new challenges such as the arising of new questions on the topic of smart homes. Indeed, a single smart home can generate hundreds of thousands transactions every day and it is not always simple to store them adequately over the long term. While it is not currently the case, we can imagine that in the future companies or government will have to manage incoming data from dozen of smart homes creating tremendous challenges and opportunities. Until now, smart homes have been conceived as standalone projects developed individually by each team of research and making them often unique in their kind. The consequence is incompatible smart homes, which prevents the development of systems using network of collaborating smart homes. Over the long term, it could refrain the widespread adoption of such technology. That is why researchers need to analyze and think about the future in which smart homes will be numerous and systems will require to interact with them all.

In this book chapter, we explore two important themes. The first one is the emerging ideas and applications related to smart homes in the context of Big Data [23]. In that part of the chapter, we review how many smart homes can be exploited as one big data warehouse. Specifically, we review the applications related to assistive smart homes, our main field of interest. We discuss how it could be used to learn more about human normal profiles and how the knowledge could be used to enhance the services. For example, the monitoring of normal human could help in the prediction of the apparition of a cognitive disease or physical disabilities through time [24]. By selecting key features such as the amount of activity, the speed of execution of tasks and the efficiency, this could be more than a vision. Over time, this could help learn more about the diseases themselves and the efficiency of the different prompting techniques. Moreover, it could help in building business intelligence tools for healthcare professionals [25]. The second theme of the chapter will cover the most important challenges smart homes researchers will be facing in the advent of Big Data. The first one is related to the data format that could or should be used [26]. In particular, should we store raw data? Low level events (e.g.: door open, motion sensor activated, etc.)? High-level knowledge (e.g.: gestures, topology, resident position, etc.) [27]? Actions that are step of plans (e.g.: Milk has been poured, oven is preheating, etc.)? We will explore advantages and disadvantages of each. The next challenge presented is about the development of data mining algorithms in the new and evolving context of Big Data. Amongst other things, the questions that arise are how to exploit algorithms when the data cannot enter into the memory? How to improve learned models without repeating the whole learning process? As the reader will see, it is a difficult topic that will be discussed for years to come. Finally, it seems also very important to encourage the inclusion of professionals from other fields of research in that discussion since it might decide of the fate of smart home in the vision of Big Data.

2 Smart Home

The smart home concept is a loosely defined idea that first emerged into science fiction novels where human beings in a distant time would cohabit with artificial intelligence with whom they communicate through simple human conversation. However, in science, the concept behind the words smart home do not necessarily refers to an artificial intelligence interacting with its residents. Indeed, the research on this topic is much broader. A smart home can be any standard house with few simple automation systems. At the most basic level, these systems could be no more than reflex agents: agents which observe the environment and react accordingly to the acquired information. For example, a thermostat agent could be a small piece of software implementing a reflex based agent's function. That agent would have a simple goal (*desired_temperature*) and its function would simply be *heat()* whenever the *current_temperature*, as observed by its sensor, is inferior to the desired temperature. A smart home constituted of such simple reflex agents can be exploited to simplify the life of its residents or to improve the comfort at home. Thus, smart homes can be exploited to reach a higher quality of life. By working together, even simple agents could be used to produce interesting results. For example, imagine if the same thermostat agent could communicate with a weather monitoring/forecasting agent and a windows/blinds manager agent. Together, if those three share information, they could work to stabilize the temperature of the house and to save energy. Let us suppose the day is predicted to be hot and sunny, then the thermostat agent could lower the heating and ask the blinds manager agent to open the blinds so the sun comes in as a natural heating.

It goes without saying that smart homes can be considerably more evolved than that. However, the challenges that limit the possibility of services and home improvement are relatively unchanged. They mostly regard the data that can be obtained to represent the environment and the information that we can expect to extract (reliably). The more information one can obtain on the state of the environment, the better the services provided can be. Nowadays, a wide range of sensing technology can be used to gather different type of data and generally at a reasonable cost. Despite the availability of data, obtaining useful information is not necessarily easy. Many researchers believe that with a better understanding and usage of the data, smart homes could be exploited to assist individuals with a reduced autonomy by recognizing the activities of daily living (ADLs), the context and the occurring problems in their actual realization (or operation) [28, 29]. In the remaining of this section, we introduce the reader to the fundamental concepts of smart homes that could lead to such an advance (cognitive and punctual assistance). The next subsection briefly describes the technology behind smart homes. Then, the section after describe the main challenges that remain to overcome on a computer science perspective. Finally, the section concludes by explaining the convergence of Big Data research with smart home and automation.

2.1 Technology

Before entering a long discussion of the major trend and challenges of smart home research, an overview of the technology is essential. First, technological design of a smart home can be divided in three main aspects: networking, sensors and effectors. There is no consensus on the exploitation of communication technologies into smart home [7]. For many years, the most extended technology was the Power Line Carrier (PLC) used with a standard protocol such as X10. This technology is useful since it enables one to exploit directly the electrical system of a house without performing any modification to it. The performance of PLC is generally very good but may be suffering from interference from few home appliances [30]. The use of PLC is still limited by the types of compatible technology and the number of devices that can use it. In smart homes, those for assistance at least, it is often necessary to have over a hundred sensors and effectors, which is not possible by relying solely on PLC. Other smart home prototypes use customized Ethernet networks [6, 22] for the communication. The sensors are then usually wired to communication hubs that can read the electrical signal and send it over Ethernet cable. The method is probably the most robust, but it may not always be possible to implement it due to the cost of modifying existing building and willingness of the residents/owners. Nowadays, the new smart homes projects mostly rely on the various wireless technologies existing. The most popular are probably WiFi, Z-Wave and ZigBee [31]. These technologies are increasingly reliable and offer more flexibility for the installation. However, sensor networks based on wireless usually require batteries, which may be viewed as an important downside. Finally, general population often view radio waves as something dangerous for health and it may refrain the adoption of smart home technology. Notes that the World Health Organization believes that it is very unlikely that radio waves affect human health [32].

On the sensors level, smart home design is divided into two opposite philosophies. A part of the literature relies on computer vision and different kinds of camera [33, 34]. The advantages of vision based system are the high quantity of information that can be extracted from only a small number of sensors. While we lacked the required processing power and the good algorithms a decade ago, it is now possible to recognize faces in real time and extract different objects from a picture with high accuracy. The main drawback is that the computer vision systems are less scalable, meaning they require a lot of configuration and a simple environmental change (wall paint, sunlight, etc.) can disrupt the system. The second philosophy in smart home is to integrate and hide a multitude of simpler and cheaper sensing technologies in the environment. Just to name a few, there are electromagnetic contacts (useful for cabinets and other doors), tactile/pressure mats (for localization), infrared motion detectors (for localization), temperature/humidity sensors (for comfort, etc.), accelerometers, flowmeters and smart energy switches (to detect plugged appliances). Other technologies are also increasingly used into smart homes to obtain more information. For example, passive RFID can be used to track objects in real-time [35], smart power analyzer to detect load signature of appliances [36] and microphones to recognize targeted activity [37].

Both approaches to the smart home field still face several challenges, and despite heated discussions, it is still unclear which will prevail over time. Finally, there is also a discussion about the effectors and actuators that can be integrated in the smart home, but it mainly depends on the application of the said smart home (comfort, energy saving, cognitive assistance, etc.) [19]. Figure 1 shows the DOMUS' smart home prototype which comprise around two hundreds sensors.

2.2 Classic Challenges

Smart home research is a broad topic requiring expertise from many disciplines such as engineering, mathematics, occupational therapy, architecture, etc. However, in computer science, few challenges have occupied a much more important part of the research effort through the years. The first one that we wanted to discuss is the difficult problem of activity recognition [38]. Activity recognition is an instance of an old and well-known problem of computer science named the plan recognition. It has been a very active topic during the past few decades [39] following the large success of expert systems, which were exploited for planning. In the context of smart home, the task is often characterized as a keyhole context [40] as opposed to intended activity recognition. In the intended case, it is assumed that the user know that he is interacting with a computer system and will adjust accordingly to help the recognition process. In the keyhole context, the recognition is performed unbeknownst to the user. Therefore, in may be assumed that the challenge can be significantly harder, especially if the user is impaired (physically or cognitively) since he might perform action that are not part of his plan (i.e.: errors).

Fig. 1 The DOMUS' smart home prototype

We can distinguish two big families of activity recognition approaches. The first one is said to be knowledge driven and regroups probabilistic and logic based algorithms [13, 41]. The first hypothesis that is made for this task is the existence of a plan structure made up by an actor agent toward the realization of one or many high-level goals. It is by acting in his environment that the observing agent will perceive information. That information on the form of raw data must be transformed into high-level actions. Then using those perceived actions, the observing agent will construct a set of plans' hypotheses from its own knowledge base. That knowledge base is assumed to contain all possible plans that can be realized by the actor. Figure 2 summarizes the relation that exists between the actor and the observer. These approaches cover usually four steps: data collection, sensors fusion, high-level action inference and activity recognition. Each of these steps are concerned with their own difficulties. The main limitation lies in the knowledge base of the observing agent which require significant expertise from a human participant mainly because it is assumed to be exhaustive and correct. Additionally, it is encoded in a complex formalism (first order logic, HMM, etc.) that only highly qualified computer science experts master.

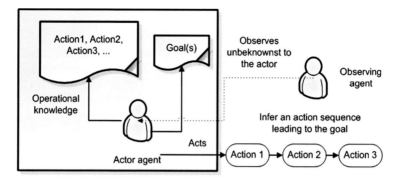

Fig. 2 Relation between an actor agent and an observing agent trying to recognize ADLs in a smart home

The second branch of approaches is said to be data driven. It has mainly emerged due to the limitations related to the knowledge library required by the knowledge driven approaches. These approaches are the most interesting for this chapter as they are the source of the emerging Big Data vision of smart homes. They are usually hybrid, but can rely purely on data mining techniques or on other machine learning methods. The challenges related to data mining are described in Sect. 4.2 of this chapter. The main problems with this part of the literature is that the approaches are mostly supervised and thus require unrealistic human efforts to construct a good dataset. While there are many successful attempts at solving the activity recognition problem on data driven approach, the advances are still embryonic with only limited experiments on small datasets (often composed of only few thousands records).

2.2.1 Context Awareness

Finally, while it may be less interesting for the purpose of this chapter, another big trend the research on smart home is what we call *context awareness*. It is often view as complementary to activity recognition since it regards mostly the understanding of the activity and the location. Context aware applications are applications that can sense the context (gather information such as the activity and the location), understand it and act accordingly. Thus, it is a desired property of the smart home to be characterized as a context aware application. The main challenge outside the information acquisition (activity recognition and localization) is the representation of the context and related knowledge in a way it can be useful for an intelligent agent. To learn more about context awareness see [42].

2.3 Emergence of Big Data

The first part of the Sect. 2 reviewed the classical visions and problems expressed in the literature on smart home research. However, the tendency in the field is to focus more and more on data driven approach leading the field to enter in the era of data. The researchers now need to start thinking about the future context were Big Data will become the standard and think about the problems and opportunities arising from it.

In the early years of smart home research, the technology was much more limited than it is nowadays. The sensors had lower sampling rates and the limited processing power restrained the smart home prototype to only keep minimal equipment. Moreover, for price and availability reasons, the sensor selection was more limited. New smart homes prototypes generate much more data than their predecessor due to those new facts. This new data could lead us to obtain more information on the context and then consequently provide much better services to the resident.

To illustrate that new context, let's take the new smart home prototype of the DOMUS laboratory. The basic sensors are collected every 200 ms and there are 221 of them with at least 150 used in most software (some are actually unused). The basic sensors include the sensors that usually do not require further processing to obtain useful data such as electromagnetic contacts, flowmeters, IR motion, temperature sensors, etc. This set of sensors alone produce 3,978,000 entries per day (221 sensors * 5 * 3600 s). The DOMUS's prototype is also equipped with 20 RFID antennas distributed in the environment. By default, the data is collected every 200 ms but the system can be speeded up tenfold for precise localization of objects. Also, each object of the smart home is tagged (with passive tags) and for precise localization it is better to put 2 or 4 tags per object (the main problem of passive RFID localization is the bad angles of arrival). Therefore, in the best case the RFID system generates 14,400,000 entries per day (40 tags * 20 antennas * 5 * 3600 s) and in the worst case 576 million (160 tags * 20 antennas * 50 * 3600 s). Finally, the DOMUS is equipped with a smart power analyzer that is plugged directly on the electrical box and read the the load signature at a sampling rate of 60 Hz resulting into 216,000 data entries per day (60 * 3600 s). These numbers are obviously a bit exaggerated as they were calculated naively, but it illustrates how a modern smart home can generate large datasets. We discuss in Sect. 4.1 the different methods that can be used to represent the data and reduce the size of the data warehouse.

Now let us put aside the DOMUS's smart home prototype and think on a larger and more realistic scale. First, the DOMUS is a standard apartment, not a full-size house. It is easy to imagine that in the future smart homes will be implemented in bigger building and thus generates more data. For example, the DOMUS laboratory has recently built ten smart homes in a center for longer care. Due to the difficulty of the project, the data is only processed in real-time and never stored on the servers. However, this data could embed important information and could be used for a whole lot of new purposes and it may be a waste to discard it like that.

Moreover, the main problem of data centric approaches to activity recognition in the literature is the lack of data. Thus collecting the data not only open the gates to new services and applications, but it also enable us to finally test data centric algorithms to activity recognition on a larger scale. Secondly, in the future, smart homes could (if we succeed) be numerous in a city and integrated in the whole concept of smart city. While it is questionable to ask whether it is a good idea or not to store it, a large set of smart homes could generates billions of data entries every day. This future could be closer than one think. Thus, we argue that researchers should start thinking about smart home in the era of Big Data.

3   Applications of Big Data

The previous section introduced the reader to the fundamental concepts related to the smart home research. It also aimed to convince the reader of the emergence of Big Data within the context of research on smart homes. Yet, the reader might be wondering if Big Data can really change research on smart home and if it is justified to think about persistence of collected data. While there is no definitive answer to that question, there are potential applications, to assistive smart home at least, that may never be realized if smart home research ignore the Big Data era. This section introduces some of these potential applications. Keep in mind that minimally, Big Data would help improve data centric methods to activity recognition. As we discussed in Sect. 2, these methods suffer from unrealistically small dataset and a very limited set of activities to be recognized. With the current advance of the literature, researchers cannot be sure that these models would scale to twenty or thirty ADLs and to a much bigger dataset (granted it is often supposed that with more data the models improve).

3.1   Learning the Efficiency of Prompts

Recently researchers have started to take a step back on assistive technologies to look at the strategies to assist the user [43]. Indeed, supposing that we can actually identify errors, assistance opportunities and dangerous situations, it is not clear what the best way to actually intervene is. Few studies has been conducted [19, 43] and they are limited to a small context. Big Data could greatly contribute to help identifying and confirming what are the good prompt methods to use. Supposing that we cumulate data from multiple smart homes over time that use different prompting modality, we could learn the efficiency of prompt by comparing similar assistance and the evolution of the ability of the resident over time. This information have significant value for the future of smart home as assistive tools and for better services delivery. Additionally, learning how the residents react to different prompts could have an impact on healthcare sciences. Indeed, the knowledge could be exploited to improve occupational therapy practices for example.

3.2 Learning About the Profile of the Resident

In addition to the learning of efficiency of prompting strategies, a Big Data warehouse built from multiple smart homes could be used to learn several things on residents and their profile. For instance, some researchers are working toward emotion recognition [44]. Emotion recognition is usually performed with video camera or other vision based sensors. It is possible that a person changes her habit and performance correspondingly to her emotion. For example, supposes that the resident is angry while preparing the diner. The motion of the objects in the environment might be faster reflecting on the current emotional turmoil. Currently, it is difficult to understand the sensors patterns corresponding to emotion, but with an important dataset, this might be possible to do. In addition to the recognition of emotion from the learning of patterns in the data warehouse, we could make the hypothesis that information about the diseases and the change of state of the resident is embedded in the data. Many diseases must be monitored closely in order to evaluate the person. For example, if the resident suffers from the Alzheimer's disease, his state will slowly worsen during a period of about a decade. Supposing that the smart home is equipped with the technology and a system to recognize the gesture (or more generally the movement) in the smart home [45], this information over time could be exploited to understand the progression of the disease and more accurately recognize the state in which the person is (it is something difficult to do for physicians due to limited information). We could even push the research further by trying to learn about normal and abnormal behaviors. By learning the patterns, it is possible that we become able to predict the cognitive disease before they appear. Obviously, these are only hypotheses for now, but with the collection of a Big Data warehouse, many might become reality in the future.

4 Challenges

In the previous sections, we discussed the emergence of Big Data in the smart home context and few of the potential applications. However, for this vision to become reality, many challenges are awaiting to be solved. In this section, we review some of the most important. Particularly, we discuss the data format, and the difficulties of the data mining process in the era of Big Data.

4.1 Data Format

There are many challenges awaiting researchers for the implementation and implantation of smart home networks. We obviously cannot pretend to cover them all so we decided to focus on few aspects that seemed more important to discuss.

The first one is the central piece of this vision of smart home in the Big Data era. Which format to use to save the data from the sensors of habitat? While this question may seems superficial, we hope to convince the reader in this section that the implications are very important and have consequence for the use of a data warehouse. Four methods that researchers have implemented are discussed along with their impacts.

4.1.1 Raw Data

Data mining and Big Data researchers might tell you that there is only one true way to build a data warehouse: storing the raw data. In the case of our future network of smart homes, that means collecting directly the data that come from the various types of sensing technologies. The idea is thus to read the value for each sensors and store it directly. There is two important advantages to this method. First, it is very simple to implement (at least on some level). There is no preprocessing of the data before the storage so it gives a degree of freedom on the side of the data mining and exploitation of the data. Moreover, reusing the data without knowing the context and the infrastructure is possible. Only the reading speed is required to be able to do so. The speed may however pose some challenges. Indeed, not all the sensing technologies can be synchronized at the same reading speed nor it is useful for all of them. For instance, an infrared motion sensor does not need to be read faster than once or twice per second, but passive RFID technology require a much higher rate to become useful (at least for localization). The second advantage of storing the raw data is the informativeness which remains intact. Since no transformation is done, the warehouse retains all the information embedded in the data. Therefore, with that format of data, there is more chances that interesting information can be found. Despite these advantages, we haven't found any case of smart homes prototype in the literature that claimed storing raw data. In reality, the problem is that this method generates a lot of data as demonstrated by the calculation in Sect. 2.3. Moreover, a large part of the data is composed of useless information and duplicates. Table 1 shows a simplified example of raw data storage.

Table 1 Example of raw data storage

Read/100 ms	Motion 1	Temperature 1	RFID 1	RFID 2	...	RFID N
#1	0	25	52	33	...	12
#2	0	25	51	29	...	14
#3	0	25	56	28		89
...	...	...	...	...	...	...
#N	1	24	49	32	...	86

4.1.2 Events Based

The second storage method is what we call events based storage. In the smart home literature, most of the data mining experiments have been conducted on events based datasets. In that paradigm, instead of storing raw data from the sensors, only the changes of state are stored. For example, if an IR motion sensor is returning a 0 (no motion is detected) and then a 1 is read, an event will be recorded in the dataset (motion detected). Events based storage is closely tied to time. Indeed, since the data is not recorded in real time, the time lapse between two entries if different. Events based storage require less space than raw data since it removes the duplicate. However, the data structures exploited is not homogeneous since not all events can be stored with the same information. The main advantages of the method is that the dataset is easy to understand and process and the risk of losing information is limited for simple sensors. However, the story is different for more complex sensing technologies such as ultrasound and laser range scanner. Definition of events is harder since the value of these sensors varies every reading. In that case, threshold must be set and it increases the risk of losing information. Moreover, even if it reduces the dataset a lot in comparison to raw data storage, the dataset will still grow very big over time. Thus, the advantage is mostly short term. Table 2 shows an example of events based storage.

4.1.3 High Level Information

A third type of storage method is the transformation of the data in high-level information. The idea is to try interpreting the raw data and extract more valuable information. For instance, instead of storing that a motion sensor has been activated (event), one could instead store the position of the resident by trying to perform localization. Obviously, the disadvantage is that it depends on the technology exploited. Moreover, there is a high risk of losing information while doing the

Table 2 Example of events based storage

Time	Sensor	State	Description
2014/10/31 12:32:23	Motion 1	Off	Resident left zone 1
2014/10/31 12:33:24	Motion 3	On	Resident in zone 3
2014/10/31 12:33:25	RFID 1	(23, 45)	Object moved in (23, 45)
2014/10/31 12:33:27	RFID 1	(27, 49)	Object moved in (27, 49)
…	…	…	…
2014/10/31 13:12:18	Temperature #1	28	Temperature increased to 28

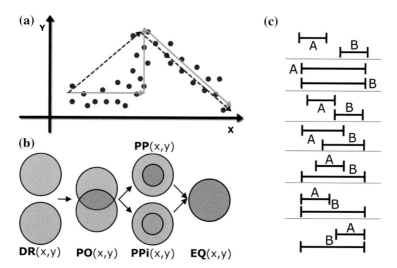

Fig. 3 **a** Extraction of qualitative directions from a dataset of positions, **b** topological relations (RCC 5 [46]) and **c** temporal relations between events

transformation. On the other side, if performed correctly, the gain in information might outpace the risk. For example, knowledge of spatial nature such as topology, gesture and orientation could be extracted. The main advantage of this method is that it reduces the size of the dataset significantly due to the transformation. However, it is not clear which information should be retained and which would be abandoned. This could have as a consequence of creating incompatibility between smart homes if the types of information is are too different. Figure 3 illustrates three possible types of qualitative information.

4.1.4   High Level Actions

The final type of information storage that we wanted to discuss is the dataset based on high-level actions. The idea of this method is to try to perform the recognition of actions that could be part of a plan or an activity of daily living. Therefore, instead of focusing on the data directly, the storage include only the significant actions of the resident. For example, if the motion detector of the living room and the smart power analyzer detect a load signature corresponding to the television, the information stored could be *open the tv* as part of the ADL *watch tv*. If other information are collected such as the temperature has risen or anything else, it is only stored if it corresponds to the step of a plan. The method has many advantages. First of all, the dataset will grow very slowly and it is possible that in fact that the warehouse remains under a reasonable size. Secondly, most of the classical activity recognition algorithms make the assumption that the basic actions are recognized. Therefore, with this storage method activity recognition should be facilitated. However, the recognition of basic actions of ADLs is a very difficult challenge. It is in fact one of the reason that prevented classical algorithms such as the one of Kautz [13] to be concretely used into smart home prototypes. Moreover, the storage of high level actions might provoke the loss of precious information. Indeed, as we justified in Sect. 3, the accumulation of a data warehouse could enable data mining researchers to find interesting information. Data mining of high-level actions is not impossible, but there is a significant chance that the discoveries would be limited.

In conclusion, as we have seen, each method possesses some advantages and some limitations. It is also possible that the best method to employ could be a hybrid between two or more of the preceding.

4.2 Data Mining Challenges

The second part of this section focuses on the challenges related to data mining in the context of Big Data. To understand these challenges, what is meant by data mining must first be defined appropriately. Data mining is the set of methods and algorithms allowing the exploration and analysis of database [47]. It exploits tools from statistics, artificial intelligence and SGBD. Data mining is used to find patterns, associations, rules or trends in datasets and usually to infer knowledge on the essential part of the information [48]. It is often seen as a subtopic of machine learning. However, machine learning is typically supervised, since the goal is to simulate the learning of known properties from *experience* (training set) in an intelligent system. Therefore, a human expert usually guides the machine in the learning phase [49]. Within realistic situations, it is often not possible to do. While the two are similar in many ways, generally, in data mining the goal is to discover previously *unknown* knowledge [25] that can then be exploited in business intelligence to make better decisions or with artificial intelligence to perform some computation (deliberation).

The complete process of data mining is illustrated on Fig. 4. Before beginning the cycle, it is important to understand the context and the data related to our situation. For example, what is the goal of the data mining? What are the consequences of errors? Are they insignificant (marketing) or critical (healthcare)? Data consideration is also important but usually for the strategy design. First of all, what types of attributes are interesting? Is there any strong association between two attributes? Those are examples of questions one should try to answer before even beginning the data mining cycle. The first step is to collect and clean the data from potentially more than one source, which can be devices, sensors, software or even websites. The goal of this step is to create the data warehouse that will be exploited for the data mining. The second step consists in the preparation of the data in the format required by the data mining algorithm. Sometime in this step, the numerical values are bounded; other time, two or more attributes can be merged together. It is also at this step that high level knowledge (temporal or spatial relationships, etc.) can be inferred for suitable algorithms. The next step is the data mining itself. It is

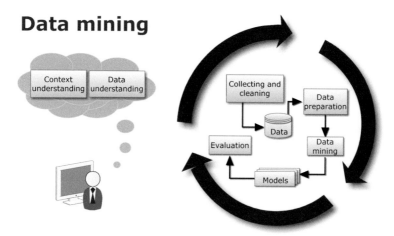

Fig. 4 The overall data mining process

important to choose or design an algorithm for the context and the data. There are many algorithms to be used. Finally, the data mining step should results in a set of models (decision trees, rules, etc.) that need to be evaluated. In a supervised context, it is usually easily done with statistical methods such as the F-Measure, K-Statistic or the ROC curve [47]. However, in an unsupervised context it is often required to design more complex validation processes. If the evaluation is not conclusive enough, the cycle can be repeated many times. Indeed, data mining is a method that often does not give expected results the first time. Note that the collection and cleaning step is generally done only once regardless of the results.

4.2.1 Supervised or Unsupervised Learning

As we discussed before, whether we talk about data mining method or machine learning in general, the process is usually classified under different categories [50]. The first one is supervised learning. The method is said supervised since it is based on training dataset with labeled examples or classes. The signification is that the algorithm can create a model that describes each class by using the known answers in the training set. In that situation, the idea is to generalize a function that maps the input to the output, and that can be used to generate output for previously unseen situations. The main implication is that somehow a human expert on the subject must label the dataset. On the opposite, unsupervised learning [49] works by using unlabeled examples. The idea is then to find hidden structure or association within the dataset and generalize a model from it. The results are sometime disappointing whether or not hidden knowledge exists in the dataset, but also sometime very surprising as the users do not know necessarily what they look for. The main implication is that there is no reward signal to evaluate the potential solutions.

Unsupervised learning is often much harder to implement. Some researchers also use the name semi-supervised learning to describe their models. In that case, it usually means that the training set is partially labeled. However, it is also used to mean that unsupervised learning was applied on a training set divided into several classes by a human or an algorithm [51].

4.2.2 Data Mining in the Context of Big Data

With the emergence of Big Data, data mining needs to evolve in order to become adapted to the new challenges that have arisen. In particular, one of the most interesting and the most difficult issue is due to the incapacity to load all the data into the Random-Access Memory (RAM) of the computer. Because of this, classical data mining algorithms do not work; they must be adapted. There are several branches of the research that try to address this problem in their own way. For example, some are working on the parallelization of the algorithms [52] in order to load all the data in the RAM of a cluster of computers. Others are trying to exploit advanced sampling method to extract representative data set from the big warehouses. However, Big Data is usually a context with low information density, which poses impossible challenges to the sampling. It is particularly the case for smart home applications. If the data is collected directly from the sensors without any transformation, there is a lot of repetition and only a small portion might be very interesting. Another possibility is to try to aggregate the low-level data into a smaller set of high level knowledge. As we discussed in Sect. 4.1, it is possible to transform the collected data into high-level knowledge, but there are few disadvantages. Moreover, it is not clear that it will prevent the dataset to grow to a large data warehouse.

In addition to the difficulties related to the memory of computer to process the data, another important question arise in the context of Big Data. Since the data warehouse is big, it might take some time to process it entirely. The classical data mining algorithms do not propose any method to revise learned models with new data. Currently, data mining process must be repeated every time that one needs to integrate new data. With Big Data warehouse, this process is long and complex, and thus it would be interesting to develop algorithms that dynamically improve learned models from new incoming data [24]. These challenges are very important and will need to be addressed in the future if smart home is to enter the era of Big Data.

5 Conclusion

In this chapter, we have tried to introduce the smart home research into the era of Big Data. In our knowledge, we are among the first to envision the future of the discipline into that particular context. As we have discussed, there are many reasons for smart home to empower the Big Data context. We have discussed some of the potential applications that could results from the collection of a Big Data warehouse. For instance, it is possible that the data warehouse contains information on the resident such as the patterns related to the progression of a disease. Then, we have discussed some of the challenges that are emerging from this new context. In particular, we have seen that the collection of data might be done under different formats. Each of these possess their advantages and disadvantages. We also discussed the data mining process and the challenges related to the Big Data context. In conclusion, we believe that many difficulties lies ahead for the smart home research, but that the potential applications are worth entering into the era of data.

Acknowledgments The authors would like to thank their mains financial sponsors: the Natural Sciences and Engineering Research Council of Canada (NSERC), the Quebec Research Fund on Nature and Technologies (FRQNT), the Canadian Foundation for Innovation (CFI).

References

1. Augusto, J.C., Nugent, C.D.: Smart homes can be smarter. In: Designing Smart Homes: Role of Artificial Intelligence, vol. 4008, pp. 1–15. Springer, Berlin (2006)
2. Weiser, M.: The computer for the 21st century. Sci. Am. **265**, 66–75 (1991)
3. Intille, S.S.: Designing a home of the future. IEEE Pervasive Comput. **1**, 76–82 (2002)
4. Robles, R.J., Kim, T-h: Applications, systems and methods in smart home technology: a review. Int. J. Adv. Sci. Technol. **15**, 37–48 (2010)
5. Nugent, C., Mulvenna, M., Moelaert, F., Bergvall-Kareborn, B., Meiland, F., Craig, D., Davies, R., Reinersmann, A., Hettinga, M., Andersson, A.-L., Droes, R.-M., Bengtsson, J.E.: Home based assistive technologies for people with mild dementia. In: Proceedings of the 5th International Conference on Smart Homes and Health Telematics, pp. 63–69. Springer, Nara (2007)
6. Giroux, S., Leblanc, T., Bouzouane, A., Bouchard, B., Pigot, H., Bauchet, J.: The praxis of cognitive assistance in smart homes. In: Gottfried, B., Aghajan, H.K. (eds.) Behaviour Monitoring and Interpretation, vol. 3, pp. 183–211. IOS Press, BMI Book, Amsterdam (2009)
7. de Vicente, A.J., Velasco, J.R., Marsa-Maestre, I., Paricio, A.: A proposal for a hardware architecture for ubiquitous computing in smart home environments. In: Proceedings of the Ist International Conference on Ubiquitous Computing: Applications, Technology and Social Issues, Alcala de Henares, Madrid, Spain, 7–9 June 2006. CEUR-WS.org (2006)
8. Hoey, J.: Tracking using flocks of features, with application to assisted handwashing. Br. Mach. Vis. Conf. BMVC **1**, 367–376 (2006)
9. Mihailidis, A., Barbenel, J.C., Fernie, G.: The efficacy of an intelligent cognitive orthosis to facilitate handwashing by persons with moderate to severe dementia. Psychology Press, Hove, Royaume-Uni (2004)
10. Mäkelä, K., Belt, S., Greenblatt, D., Häkkilä, J.: Mobile interaction with visual and RFID tags: a field study on user perceptions. In: Proceedings of the SIGCHI Conference on Human Factors in Computing Systems, pp. 991–994. ACM, San Jose, California, USA (2007)
11. Helal, S., Mann, W., El-Zabadani, H., King, J., Kaddoura, Y., Jansen, E.: The gator tech smart house: a programmable pervasive space. Computer **38**, 50–60 (2005)
12. Cook, D.J., Youngblood, M., Heierman, I, E.O., Gopalratnam, K., Rao, S., Litvin, A., Khawaja, F.: MavHome: an agent-based smart home. In: Proceedings of the First IEEE International Conference on Pervasive Computing and Communications, pp. 521–524. IEEE Computer Society (2003)

13. Kautz, H.A.: A formal theory of plan recognition and its implementation. Reasoning about plans, pp. 69–124. Morgan Kaufmann Publishers Inc. (1991)
14. Roy, P., Bouchard, B., Bouzouane, A., Giroux, S.: A hybrid plan recognition model for Alzheimer's patients: interleaved-erroneous dilemma. Web Intelli. Agent Syst. **7**, 375–397 (2009)
15. Charniak, E., Goldman, R.P.: A Bayesian model of plan recognition. Artif. Intell. **64**, 53–79 (1993)
16. Nguyen, N., Bui, H., Venkatesh, S., West, G.: Recognising and monitoring high-level behaviours in complex spatial environments. In: IEEE International Conference on Computer Vision and Pattern Recognition (CVPR), pp. 620–625 (2003)
17. Palmes, P., Pung, H.K., Gu, T., Xue, W., Chen, S.: Object relevance weight pattern mining for activity recognition and segmentation. Pervasive Mob. Comput. **6**, 43–57 (2010)
18. Gu, T., Wang, L., Wu, Z., Tao, X., Lu, J.: A pattern mining approach to sensor-based human activity recognition. IEEE Trans. Knowl. Data Eng. **23**, 1359–1372 (2011)
19. Lapointe, J., Bouchard, B., Bouchard, J., Bouzouane, A.: Smart homes for people with alzheimer's disease: adapting prompting strategies to the patient's cognitive profile. In: Proceedings of the 5th ACM International Conference on Pervasive Technologies Related to Assistive Environments. ACM publisher, Crete Island, Greece (2012)
20. Gouin-Vallerand, C., Abdulrazak, B., Giroux, S., Mokhtari, M.: A self-configuration middleware for smart spaces. Self **3** (2009)
21. Ramos, C., Augusto, J.C., Shapiro, D.: Ambient intelligence: the next step for artificial intelligence. IEEE Intell. Syst. **23**, 15–18 (2008)
22. Bouchard, K., Bouchard, B., Bouzouane, A.: Guideline to efficient smart home design for rapid AI prototyping: a case study. In: International Conference on Pervasive Technologies Related to Assistive Environments. ACM (2012)
23. Frankel, F., Reid, R.: Big data: distilling meaning from data. Nature **455**, 30–30 (2008)
24. Ordonez, C.: Can we analyze big data inside a DBMS? In: Proceedings of the Sixteenth International Workshop on Data Warehousing and OLAP, pp. 85–92. ACM (2013)
25. Chaudhuri, S., Dayal, U., Narasayya, V.: An overview of business intelligence technology. Commun. ACM **54**, 88–98 (2011)
26. Coyle, L., Neely, S., Stevenson, G., Sullivan, M., Dobson, S., Nixon, P.: Sensor fusion-based middleware for smart homes. Int. J. Assist. Robot. Mechatron. (IJARM) **8**, 53–60 (2007)
27. Asadzadeh, P., Kulik, L., Tanin, E.: Gesture recognition using RFID technology. Pers. Ubiquit. Comput. **16**, 225–234 (2012)
28. Rashidi, P., Cook, D.J., Holder, L.B., Schmitter-Edgecombe, M.: Discovering activities to recognize and track in a smart environment. IEEE Trans. Knowl. Data Eng. **23**, 527–539 (2011)
29. Chen, L., Nugent, C.D., Wang, H.: A knowledge-driven approach to activity recognition in smart homes. IEEE Trans. Knowl. Data Eng. **24**, 961–974 (2012)
30. Murty, R., Padhye, J., Chandra, R., Chowdhury, A.R., Welsh, M.: Characterizing the end-to-end performance of indoor powerline networks. Harvard University Microsoft Research (2008)
31. Gomez, C., Paradells, J.: Wireless home automation networks: a survey of architectures and technologies. IEEE Commun. Mag. **48**, 92–101 (2010)
32. Valberg, P.A., van Deventer, T.E., Repacholi, M.H.: Workgroup report: base stations and wireless networks-radiofrequency (RF) exposures and health consequences (2007)
33. Hoey, J., Poupart, P., Bertoldi, Av, Craig, T., Boutilier, C., Mihailidis, A.: Automated handwashing assistance for persons with dementia using video and a partially observable Markov decision process. Comput. Vis. Image Underst. **114**, 503–519 (2010)
34. Nguyen, N.T., Phung, D.Q., Venkatesh, S., Bui, H.: Learning and detecting activities from movement trajectories using the hierarchical hidden markov models. In: Proceedings of the 2005 IEEE Computer Society Conference on Computer Vision and Pattern Recognition (CVPR'05), vol. 2, pp. 955–960. IEEE Computer Society (2005)

35. Fortin-Simard, D., Bouchard, K., Gaboury, S., Bouchard, B., Bouzouane, A.: Accurate passive RFID localization system for smart homes. In: Proceedings of the 3rd IEEE International Conference on Networked Embedded Systems for Every Application. IEEE, Liverpool, UK (2012)
36. Belley, C., Gaboury, S., Bouchard, B., Bouzouane, A.: Efficient and inexpensive method for activity recognition within a smart home based on load signatures of appliances. J. Pervasive Mob. Comput. 1–20 (2013)
37. Chen, J., Kam, A.H., Zhang, J., Liu, N., Shue, L.: Bathroom activity monitoring based on sound. Pervasive Comput. 47–61 (2005) (Springer)
38. Schmidt, C.F., Sridharan, N.S., Goodson, J.L.: The plan recognition problem: an intersection of psychology and artificial intelligence. Artif. Intell. 11, 45–83 (1978)
39. Carberry, S.: Techniques for plan recognition. User Model. User-Adap. Inter. 11, 31–48 (2001)
40. Cohen, P.R., Perrault, C.R., Allen, J.F., Agency, U.S.A.R.P.: Beyond question-answering. Bolt Beranek and Newman Inc. (1981)
41. Wobcke, W.: Two logical theories of plan recognition. J. Logic Comput. 12, 371–412 (2002)
42. Abowd, G.D., Dey, A.K., Brown, P.J., Davies, N., Smith, M., Steggles, P.: Towards a better understanding of context and context-awareness. In: Handheld and Ubiquitous Computing, pp. 304–307. Springer, Heidelberg (1999)
43. Sohlberg, M.M., Fickas, S., Hung, P.-F., Fortier, A.: A comparison of four prompt modes for route finding for community travellers with severe cognitive impairments. Brain Inj. 21, 531–538 (2007)
44. Lee, C.-C., Mower, E., Busso, C., Lee, S., Narayanan, S.: Emotion recognition using a hierarchical binary decision tree approach. Speech Commun. 53, 1162–1171 (2011)
45. Bouchard, K., Bouchard, B., Bouzouane, A.: Gesture recognition in smart home using passive RFID technology. In: Proceedings of the 7th International Conference on Pervasive Technologies Related to Assistive Environments. ACM (2014)
46. Cohn, A.G., Bennett, B., Gooday, J., Gotts, N.M.: Qualitative spatial representation and reasoning with the region connection calculus. Geoinformatica 1, 275–316 (1997)
47. Witten, I.H., Franck, E.: Data mining: practical machine learning tools and techniques. Elsevier editor (2010)
48. Quinlan, J.R., Ghosh, J.: Top 10 algorithms in data mining. In: International Conference on Data Mining (ICDM'06). IEEE (2006)
49. Barlow, H.B.: Unsupervised learning. Neural Comput. 1, 295–311 (1989)
50. Carbonell, J., Michalski, R., Mitchell, T.: An overview of machine learning. In: Michalski, R., Carbonell, J., Mitchell, T. (eds.) Machine Learning, pp. 3–23. Springer, Berlin (1983)
51. Jakkula, V., Cook, D.J.: Mining sensor data in smart environment for temporal activity prediction. In: KDD'07. ACM (2007)
52. Li, J., Liu, Y., Liao, W.-k., Choudhary, A.: Parallel data mining algorithms for association rules and clustering. In: International Conference on Management of Data. Citeseer (2008)

Authors Biography

Kevin Bouchard postdoctoral fellow at the Université de Sherbrooke. He obtained his Ph.D. from the Université du Québec à Chicoutimi (UQAC) in 2014. His research, primarily focused on activity recognition in smart home, spatial reasoning, data mining and learning, have valued him to win many prestigious scholarships such as the Alexander Graham Bell Canada Graduate Scholarship from the Natural Sciences and Engineering Research Council of Canada (NSERC).

Sébastien Gaboury is a professor and a scientist working at the Department of Mathematics and Computer Science of the Université du Québec à Chicoutimi (UQAC). He received a Ph. D. in Mathematics from the Royal Military College of Canada in 2012. His research is primary focussed on fractional calculus and special functions as well as number theory. He contributed in the domains of fractional calculus, special functions, generating functions, number theory, activity recognition and load monitoring of electrical devices.

Bruno Bouchard is an associate professor and a scientist working at the Department of Mathematics and Computer Science of the Université du Québec à Chicoutimi (UQAC). He received a Ph.D. in computer science from the University of Sherbrooke (Canada) in 2006. He then completed a postdoctoral fellowship at the University of Toronto in 2007. He co-founded in 2008, the LIARA laboratory. His research is sponsored by the NSERC, FQRNT, the Canadian Foundation for Innovation (CFI), Bell Canada, UQAC and its foundation. He contributed to application domains as varied as cognitive assistance, serious games for elders and activity recognition.

Abdenour Bouzouane is a professor at the Université du Québec à Chicoutimi since 1997. He received in 1993 a Ph.D. in computer engineering from the Ecole Centrale de Lyon. His research is funded by NSERC and FQRNT of CANADA and related to the data mining domain. He is co-founder of Ambient Intelligence Laboratory for the Recognition of activities (LIARA).

Sylvain Giroux is a professor at the University of Sherbrooke. He obtained his Ph.D. in computer science from the University of Montreal in 1993. He is currently leading the DOMUS laboratory. His research interest include assistive technologies, smart homes, context awareness and ubiquitous computing.

An Investigation of the Use of Innovative Biology-Based Computational Intelligence in Ubiquitous Robotics Systems: Data Mining Perspective

Bo Xing

Abstract Sensor technologies are crucial in ambient assisted living system because they can observe, measure, and detect users' daily activities, in the meanwhile issue warnings when parameters exceed particular thresholds. Despite their enormous supporting potential, the quantities of data generated from multiple sensor is huge (i.e., big data), because they involved everywhere, such as smart home, health-based wearable devices, and assistive robots. Generally speaking, big data can be defined as large pools of data which comes from digital pictures, videos, intelligent sensors, posts to social media sites, purchase transaction records, cell phone global positioning system signals, to name a few. During the past few years, there is a great interest both in the commercial and in the research communities around big data. Under these circumstances, data mining, whose main purpose is to extract value from mountains of datasets, is drawing a lot of people's attention. To follow this trend, in this article, we intend to take an algorithmic point of view, i.e., applying intelligent algorithms to data, with an emphasis on the biology-based innovative computational intelligence (CI) methods. This work makes several contributions. First, it investigates a set of biology-based innovative CI algorithms which can enable a high throughput under extract from the insights of data. Second, it summarizes the core working principles of these algorithms systematically and highlights their preliminary applications in different areas of data mining. This will allow us to clearly pinpoint the intrinsic strengths of these novel algorithms, and also to define the potential further research directions. The findings of this chapter should provide useful insights into the current big data literature and be a good source for anyone who is interested in the application of CI approaches to big data and its corresponding fields.

B. Xing (✉)
Computational Intelligence, Robotics, and Cybernetics for Leveraging
E-Future (CIRCLE), Department of Computer Science, School of Mathematical
and Computer Sciences, Faculty of Science and Agriculture, University of Limpopo,
Private Bag X1106, Sovenga, Limpopo 0727, South Africa
e-mail: bxing2009@gmail.com

© Springer International Publishing Switzerland 2016
K.K. Ravulakollu et al. (eds.), *Trends in Ambient Intelligent Systems*,
Studies in Computational Intelligence 633, DOI 10.1007/978-3-319-30184-6_6

139

Keywords Innovative computational intelligence · Ubiquitous robotics systems · Data mining · Artificial fish swarm algorithm · Dove swarm optimization · Firefly algorithm · Fireworks optimization algorithm · Flockbyleader · Flocking-based algorithm · Fruit fly optimization algorithm · Glowworm swarm optimization · Harmony search · Human group formation · Photosynthetic algorithm · Shark-Search algorithm · Stem cells optimization algorithm · Wasp optimization algorithm

1 Introduction

The amount of data in our world has been exploding, in particular, with the emergence of the advanced infrastructures, e.g., ambient intelligent system, Internet of things (IoT) and cloud computing. For instance, in ubiquitous robotics system domain, based on the IoT concept, the living (patients, medical staff, etc.) and non-living (medical sensors, smart devices, etc.) entities can be easily linked and thus open new opportunities for right care. Obviously, that is good which could bring great benefits to the medical staff and patients. But how we measure the value of those big data (i.e., achieving highly scalable data analysis) is a circuital challenge, as the overall applications' potential value is highly dependent on the data discovery.

The aim of this chapter is twofold: first, identifying a set of novel biology-based computational intelligence (CI) algorithms that have been applied to various data mining problems and summarizing their corresponding core working principles; second, presenting a detailed analysis regarding each of these algorithms' preliminary applications in terms of data mining. Having this in mind, the remainder of this chapter is organized as follows: in Sect. 2, the background information is briefly introduced; the core working principles of biology-based CI algorithms and their corresponding applications in data mining are detailed in Sects. 3, 4 discusses some findings obtained via this study; the limitations of the present work are outlined in Sect. 5; finally, the conclusion drawn in Sect. 6 close this study.

2 Background

2.1 What Is Big Data?

Big data refers to datasets whose size is beyond the ability of traditional or commonly used techniques to capture, store, manage, and analyze within a "tolerable elapsed time" [1]. Of course, this definition is relative and evolves in time as technology progresses. However, the question is what this phenomenon means. Is the proliferation of data simply create potential value for us? The answer is not. A shortage of the analytical and managerial method necessary to make the most of big data is a significant and pressing challenge. In addition, several authors (e.g., [2, 3]) pointed out that the focus not only on size but on three different dimensions of growth for data, i.e., volume, variety and velocity. In fact, it is not just the size of an individual data set, but rather the collection of data that is available to us through different formats (such as pictures, sound, movies, documents, and experimental measurements) and different sources (such as radio-frequency identification (RFID), GPS, and online). In a similar vein, the authors of [4] proposed HACE theorem (i.e., heterogeneous and autonomous sources, and complex and evolving relationships among data) in terms of big data characteristics. As a result, the big data forces us to use or create innovative methodologies.

2.2 Data Issue Associated with Ubiquitous Robotics Systems

Inevitably, ubiquitous robotics systems belong to data-intensive domain, due to the fact that they are often related to a large number of continuously changing data objects, such as data streams from sensors and tracking devices. As a result, the complexity of managing different sources of sensor-based information can easily make our head spin. In the light of this statement, the authors of [5] provided an insight into the use of sensor-based technology to support ambient assisted living. Additionally, in Ref. [6], a comprehensive framework for managing continuously changing data objects is presented. Indeed, sensor-based ambient intelligent system can function proactively if the information that hidden in these piles of data can be extracted.

2.3 Data Mining

But why are we interested in data? It is common belief that data without a model is just noise. In other words, data needs to be processed and condensed into more connected forms in order to be useful for decision making. That is the job of data mining. Let us now more precisely define data mining. Data mining is used to describe salient features in the data [7, 8]. For big data, it is the application of data mining techniques to discover patterns from the big amount of data.

According to the literature (e.g., [9–11]), data mining techniques can be categorized into several different types: generalization, characterization, classification, clustering, association, evolution, pattern matching, data visualization, and meta-rule guided mining. In addition, the authors of [12] argued that using a class of algorithms can effectively extract information from huge amounts of data in many data repositories or in dynamic data streams.

2.4 Innovative Computational Intelligence

Computational intelligence (CI) is a fairly new research discipline, and thus, there is little agreement regarding its accurate definition. However, most academicians and practitioners would include techniques such as artificial neural networks (ANN), fuzzy systems (FS), many versions of evolutionary algorithms (EA) (e.g., evolution strategies (ES), genetic algorithm (GA), genetic programming (GP), differential evolution (DE)), as well as ant colony optimization (ACO), artificial immune systems (AIS), multi-agent systems (MAS), particle swarm optimization (PSO), and the hybridization versions of these, under the umbrella of CI. According to [13], these algorithms is often referred to as the traditional CI.

In data mining field, ANN and GA are examples of some widely used data mining algorithms. For example, for extracting the rules form databases, the authors of [14] presented a new algorithm that via trained NN using a GA. More recently, other CI methods, such as ACO, PSO, AIS have been applied to the data mining domain. For example, the authors of [15] used ant colony decision rule algorithm for finding if-then rules for classification tasks. Also, based centrally on ACO algorithm, the authors of [16] developed a new algorithm for data clustering. Inspired by one theoretical model of the immune system, i.e., immune networks, the authors of [17] represented a new algorithm called aiNet for data clustering. In addition, [18] proposed a multi-objective chaotic PSO method as search strategy to deal with classification rule mining problem.

Although the applications of traditional CI approaches in the field of data mining is well documented by several excellent reviews (e.g., [7, 9, 19–21]), the author of this study still intends to look at data mining through CI but via another lens, i.e., innovative CI. For the sake of clarity, the algorithms covered by this work are listed as below:

- Artificial Fish Swarm Algorithm
- Dove Swarm Optimization
- Firefly Algorithm
- Fireworks Optimization Algorithm
- FlockbyLeader
- Flocking-based Algorithm
- Fruit Fly Optimization Algorithm
- Glowworm Swarm Optimization
- Harmony Search
- Human Group Formation
- Photosynthetic Algorithm
- Shark-Search Algorithm
- Stem Cells Optimization Algorithm
- Wasp Optimization Algorithm.

3 Biology-Based Innovative CI Algorithms

Briefly, biology can be defined as a comprehensive science concerning all functions of living systems [22]. From an evolutionary process point of view, biological systems possess many appealing characteristics such as sophistication, robustness, and adaptability [23]. These features represent a strong motivation for imitating the mechanisms of natural evolution in an attempt to create CI algorithms with merits comparable to those of biological systems.

3.1 Artificial Fish School Algorithm (AFSA)

Artificial fish school algorithm (AFSA), which was proposed in [24], is a stochastic search optimization algorithm inspired by the natural social behaviour of fish schooling. In principle, AFSA is started first in a set of random generated potential solutions, and then performs the search for the optimum one interactively [25]. The main steps of AFSA are outlined as follows [24, 26, 27]:

- Assuming in an n–dimensional searching space, there is a group composed of K articles of artificial fish (AF).
- Situation of each individual AF can be expressed as vector $X = (x_1, x_2, \ldots, x_k)$ is denoted the current state of AF, where $x_k (k = 1, 2, \ldots, k)$ is control variable.
- $Y = f(X)$ is the fitness or objective function of X, which can represent food concentration (FC) of AF in the current position.
- $d_{ij} = \|X_i - X_j\|$ is denoted the Euclidean distance between fishes.
- *Visual* and *Step* are denoted respectively the visual distance of AF and the distance that AF can move for each step.
- $X_v = (x_1^v, x_2^v, \ldots, x_k^v)$ is the visual position at some moment. If the state at the visual position is better than the current state, it goes forward ad step in this direction, and arrives the X_{next} state, otherwise, continues an inspecting tour in the vision.
- *try-number* is attempt times in the behaviour of prey.
- δ is the condition of jamming $(0 < \delta < 1)$.

The basic behaviours of AF inside water are defined as follows [24, 26, 27]:

- Chasing trail behaviour (AF_Follow): When a fish finds the food dangling quickly after a fish, or a group of fishes, in the swarm that discovered food. If $Y_j > Y_i$ and $\frac{n_f}{n} < \delta$, then the AF_Follow behaviour is defined as follows [24, 26–28]:

$$X_i^{(t+1)} = X_i^{(t)} + \frac{X_j - X_i^{(t)}}{\left\|X_j - X_i^{(t)}\right\|} \cdot Step \cdot rand(\ ). \tag{1}$$

- Gathering behaviour (AF_Swarm): In order to survive and avoid hazards, the fish will naturally assemble in groups. There are three rules while fish gathering: firstly, a fish will try to keep a certain distance with each other to avoid crowding (i.e., Compartmentation Rule); secondly, a fish will try to move in a similar direction with its surrounding partners (i.e., Unification Rule); finally, a fish will try to move to the centre of its surrounding partners (i.e., Cohesion Rule). If $Y_c > Y_i$ and $\frac{n_f}{n} < \delta$, then the AF_Swarm behaviour is defined as follows [24, 26–28]:

$$X_i^{(t+1)} = X_i^{(t)} + \frac{X_c - X_i^{(t)}}{\left\| X_c - X_i^{(t)} \right\|} \cdot Step \cdot rand(\ ). \tag{2}$$

where X_c denotes the centre position of AF, X_i be the AF current state, n_f be the number of its companions in the current neighbourhood ($d_{ij} < Visual$), and n is the total fish number.

- Random searching behaviour (AF_Random): This is a basic biological behaviour that tents to the food. Generally the fish perceives the concentration of food in water to determine the movement by vision or sense and then chooses the tendency. The effect of this behaviour is similar to that of mutation operator in GA (i.e., genetic algorithm). It is defined as follows [28]:

$$X_i^{(t+1)} = X_i^{(t)} + Visual \cdot rand(\ ). \tag{3}$$

- Leaping behaviour (AF_Leap): When a fish 'stagnates' in a region, it looks for food in other regions defining the leaping behaviour. It is can be defined as follows [28]:

$$\begin{cases} \text{if } (FC_{best}(m) - FC_{best}(n)) < eps \\ \text{then } X_{some}^{(t+1)} = X_{some}^{(t)} + \beta \cdot Visual \cdot rand(\ ) \end{cases}. \tag{4}$$

- Foraging behaviour (AF_Prey): As feeding the fish, they will gradually move to the place where food is increasing. It is defined as follows, respectively [24, 26–28]:

$$X_j = X_i + Visual \cdot rand(\ ). \tag{5}$$

where X_i be the AF current state and select a state X_j randomly in its visual distance, $rand(\ )$ is a random function in the range, and $Visual$ represents the visual distance.

$$\begin{cases} \text{if } Y_i < Y_j, \quad \text{then } X_i^{(t+1)} = X_i^{(t)} + \frac{X_j - X_i^{(t)}}{\left\| X_j - X_i^{(t)} \right\|} \cdot Step \cdot rand(\ ) \\ \text{if } Y_i > Y_j, \quad \text{then} \begin{cases} X_j = X_i + Visual \cdot rand(\ ) \\ X_i^{(t+1)} = X_i^{(t)} + Visual \cdot rand(\ ) \end{cases} \end{cases}. \tag{6}$$

where Y is the food concentration (objective function value), $X_i^{(t+1)}$ represents the AF's next state, and *Step* denotes the distance that AF can move for each step.

Taking into account the above mentioned behaviours, the steps of implementing AFSA can be summarized as follows [28]:

- Step 1: Generate the initial fish swarm randomly in the search space.
- Step 2: Initialize the parameters.
- Step 3: Evaluate the fitness value of each AF.
- Step 4: Selecting behaviour. Each AF simulate the swarming and following behaviour, respectively, and select the best behaviour to perform by comparing the function values, the default is searching food behaviour.
- Step 5: Update the function value of the AF again.
- Step 6: Check the termination condition.

Data Mining using AFSA. Research, such as [29], employed AFSA as new tool to discover classification rules from data. Other than being a classifier (i.e., for higher classification accuracy), the AFSA is able to automatically mine a set of smaller IF-THEN rule. The proposed method has been tested on publicly available databases (i.e., Iris Plants Database), compared with other algorithms, the computational results showed that AFSA is very effective and robust. In addition, several authors, such as [30–32], proposed a hybrid clustering method, based on AFSA and K-means. Also, the fuzzy clustering method (i.e., fuzzy C-means) is combined within AFSA, e.g., [33–35]. Simulation results showed that AFSA could achieve good clustering effects.

3.2 Dove Swarm Optimization (DSO) Algorithm

Dove swarm optimization (DSO) algorithm was recently proposed in [36]. The basic working principles of DSO are listed as follows [36]:

- Step 1: Initializing the number of doves and deploying the doves on the 2-dimensional artificial ground.
- Step 2: Setting the number of epochs ($e = 0$), and the degree of satiety, $f_j^e = 0$ for $j = 1, \ldots, M \times N$. Initializing the multi-dimensional sense organ vector, $\vec{w}_j$ for $j = 1, \ldots, M \times N$.
- Step 3: Computing the total amount of the satiety degrees in the flock, $T(e) = \sum_{j=1}^{M \times N} f_j^e$.
- Step 4: Presenting an input pattern (i.e., piece of artificial crumb) $\vec{x}_k$ to the $M \times N$ doves.

- Step 5: Locating the dove b_f closest to the crumb $\vec{x}_k$ according to the minimum-distance criterion shown below [36]:

$$b_f = \arg \min_j \left\| \vec{x}_k - \vec{w}_j(k) \right\|, \text{ for } j = 1, \ldots, M \times N. \tag{7}$$

The dove with the artificial sense organ vector which is the most similar to the artificial crumb, $\vec{x}_k$, is claimed to be the winner.

- Step 6: Updating each dove's satiety degree as follows [36]:

$$f_j^e(new) = \frac{\left\| \vec{x}_k - \vec{w}_{b_f}(k) \right\|}{\left\| \vec{x}_k - \vec{w}_j(k) \right\|} + \lambda f_j^e(old), \text{ for } j = 1, \ldots, M \times N. \tag{8}$$

- Step 7: Selecting the dove, b_f, with the highest satiety degree based on the following criterion expressed as follows [36]:

$$b_s = \arg \max_{1 \leq j \leq M \times N} f_j^e. \tag{9}$$

- Step 8: Updating the sense organ vectors and the position vectors via the following equations, respectively [36]:

$$\vec{w}_j(k+1) = \begin{cases} \vec{w}_{b_f}(k) + \eta_w \left(\vec{x}_k - \vec{w}_{b_f}(k) \right) & \text{for } j = b_f \\ \vec{w}_j(k) & \text{for } j \neq b_f \end{cases}. \tag{10}$$

$$\vec{p}_j(k+1) = \vec{p}_j(k) + \eta_p \beta \left(\vec{p}_{b_s}(k) - \vec{p}_j(k) \right), \text{ for } j = 1, \ldots, M \times N. \tag{11}$$

- Step 9: Returning to Step 4 until all patterns are processes.
- Step 10: Stopping the whole training procedure if the following criterion is met [36]:

$$\left| \sum_{j=1}^{M \times N} f_j^e - T(e) \right| \leq \varepsilon. \tag{12}$$

Otherwise, increasing the number of epochs by one ($e = e + 1$), and go back to Step 3 until the pre-defined limit for the number of epochs is met. The satisfaction of the criterion given above means that the total amount of satiety degree has converged to some extent.

Data Mining using DSO. In general there are two main obstacles encountered in data clustering: the geometric shapes of the clusters are full of variability, and the cluster numbers are not often known a priori. In order to determine the optimal number of clusters, the authors of [36] employed DSO to perform data projection task, i.e., projecting high-dimensional data onto a low-dimensional space to facilitate visual inspection of the data. This process allows us to visualize high-dimensional data as a 2-dimensional scatter plot.

The basic idea in their work can be described as follows [36]: In a data set, each data pattern, $\vec{x}$, is regarded as a piece of artificial crumb and these artificial crumbs (i.e., data patterns) will be sequentially tossed to a flock of doves on a two-dimensional artificial ground. The flock of doves adjusts its physical movements to seek these artificial crumbs. Individual members of the flock can profit from discoveries of all of the other members of the flock during the foraging procedure because an individual is usually influenced by the success of the best individual of the flock and thus has a desire to imitate the behaviour of the best individual. Gradually, the flock of the doves will be divided into several groups based on the distributions of the artificial crumbs. Those formed groups will naturally correspond to the hidden data structure in the data set. By viewing the distributions of the doves on the 2-dimensional artificial ground, one may quickly find out the number of clusters inherent in the data set.

However, many practical data sets have high-dimensional data points. Therefore, the aforementioned idea has to be generalized so that it can process high-dimensional data. In the real world, each dove has a pair of eyes to find out where crumbs are, but in the artificial world, a virtual dove does not have the capability to perceive a piece of multi-dimensional artificial crumb that is located around it. In order to cope with issue, the authors of [36] equipped each dove with functionalities, i.e., a multi-dimensional artificial sense organ represented as a sense organ vector, $\vec{w}$, which has the same dimensionality as a data pattern, $\vec{x}$, and a 2-dimensional position vector, $\vec{p}$, which represents its position on the 2-dimensional artificial ground. In addition to these two vectors, $\vec{w}$ and $\vec{p}$, a parameter called the satiety parameter is also attached to each dove. This special parameter endows a dove with the ability of expressing its present satiety status with respect to the food, that is, a dove with a low degree of satiety will have a strong desire to change its present foraging policy and be more willing to imitate the behaviour of the dove which performs the best among the flock.

To test the performance of DSO, five (two artificial and three real) data sets were selected in the study. These data sets include Two-Ellipse, Chromosomes, Iris, Breast Cancer, and 20-Dimensional Non-Overlapping. The projection capability of DSO was compared with the other popular projection algorithms, e.g., Sammon's algorithm. For DSO, the maximum number of epochs for every data set (excluding Iris and 20-Dimensional data sets) were set to be 5, while for the Iris and 20-Dimensional data sets, were set to be 10 and 20, respectively. The case studies showed that DSO can fulfil the projection task. Meanwhile, the performance of DSO is not so sensitive to the size of dove swarm.

3.3 Firefly Algorithm (FA)

Firefly algorithm (FA) is a nature-inspired, optimization algorithm which is based on the social (flashing) behaviour of fireflies, or lighting bugs, in the summer sky in the tropical temperature regions [37–39]. In FA, physical entities (fireflies) are randomly distributed in the search space. They carry a bio-luminescence quality, called luciferin, as a signal to communicate with other fireflies, especially to prey attractions [40].

In detail, each firefly is attracted by the brighter glow of other neighbouring fireflies. The attractiveness decreases as their distance increases. If there is no brighter one than a particular firefly, it will move randomly. Its main merit is the fact that the FA uses mainly real random numbers and is based on the global communication among the swarming particles (i.e., the fireflies), and as a result, it seems more effective in multi-objective optimization.

Normally, FA uses the following three idealized rules [37] to simplify its search process to achieve an optimal solution:

- Fireflies are unisex so that one firefly will be attracted to other fireflies regardless of their sex, that means no mutation operation will be done to alter the attractiveness fireflies have for each other;
- The sharing of information or food between the fireflies is proportional to the attractiveness that increases with a decreasing Cartesian or Euclidean distance between them due to the fact that the air absorbs light. Thus for any two flashing fireflies, the less bright one will move towards the brighter one. If there is no brighter one than a particular firefly, it will move randomly; and
- The brightness of a firefly is determined by the landscape of the objective function. For the maximization problems, the light intensity is proportional to the value of the objective function.

Furthermore, there are two important issues in the FA that are the variation of light intensity or brightness and formulation of attractiveness. Yang [38] simplifies a firefly's attractiveness β (determined by its brightness I) which in turn is associated with the encoded objective function. As light intensity and thus attractiveness decreases as their distance from the source increases, the variations of light intensity and attractiveness should be monotonically decreasing functions.

- Variation of light intensity: Suppose that there exists a swarm of n fireflies, and x_i, $i = 1, 2, \ldots, n$ represents a solution for a firefly i initially positioned randomly in the space, whereas $f(x_i)$ denotes its fitness value. In the simplest form, the light intensity $I(r)$ varies with the distance r monotonically and exponentially. That is determined as follows [37–39]:

$$I = I_0 e^{-\gamma r_{ij}}. \tag{13}$$

where I_0 is the original light intensity, γ is the light absorption coefficient, and r is the distance between firefly i and firefly j at x_i and x_j as Cartesian distance $r_{ij} = \|x_i - x_j\| = \sqrt{\sum_{k=1}^{d} (x_{i,k} - x_{j,k})^2}$ or the ℓ_2-norm, where $x_{i,k}$ is the kth component of the spatial coordinate x_i of the ith firefly and d is the number of dimensions we have, for $d = 2$, we have $r_{ij} = \sqrt{(x_i - x_j)^2 + (y_i - y_j)^2}$.

- Movement toward attractive firefly: A firefly attractiveness is proportional to the light intensity seen by adjacent fireflies [38]. Each firefly has its distinctive attractiveness β which implies how strong it attracts other members of the swarm. However, the attractiveness is relative; it will vary with the distance between two fireflies. The attractiveness function $\beta(r)$ of the firefly is determined via the following equation [37–39]:

$$\beta = \beta_0 e^{-\gamma r_{ij}^2}. \tag{14}$$

where β_0 is the attractiveness at $r = 0$, and γ is the light absorption coefficient which controls the decrease of the light intensity.

The movement of a firefly i at location x_i attracted to another more attractive (brighter) firefly j at location x_j is determined as follows [37–39]:

$$x_i(t+1) = x_i(t) + \beta_0 e^{-\gamma r_{ij}^2}(x_j - x_i) + \alpha \varepsilon_i. \tag{15}$$

where the first term is the current position of a firefly, the second term is used for considering a firefly's attractiveness to light intensity seen by adjacent fireflies, and the third term is randomization with the vector of random variables ε_i being drawn from a Gaussian distribution, in case there are not any brighter ones. In addition, the coefficient α is a randomization parameter determined by the problem of interest.

- Special cases: From the equation above, it is easy to see that there exit two limit cases when γ is small or large, respectively [37–39]. When γ tends to zero, the attractiveness and brightness are constant $\beta = \beta_0$ which means the light intensity does not decrease as the distance r between two fireflies increases. Therefore, a firefly can be seen by all other fireflies, a single local or global optimum can be easily reached. This limiting case corresponds to the standard particle swarm optimization algorithm. On the other hand, when γ is very large, then the attractiveness (and thus brightness) decreases dramatically, and all fireflies are short-sighted or equivalently fly in a deep foggy sky. This means that all fireflies move almost randomly, which corresponds to a random search technique.

In general, the FA corresponds to the situation between these two limit cases, and it is thus possible to fine-tune these parameters, so that FA can find the global optima as well as all the local optima simultaneously in a very effective manner. A further advantage of FA is that different fireflies will work almost independently, it is thus particular suitable for parallel implementation. It is even better than genetic algorithm and particle swarm optimization because fireflies aggregate more closely around each optimum. It can be anticipated that the interactions between different sub-regions are minimal in parallel implementation.

Overall, taking into account the basic information described above, the steps of implementing FA can be summarized as follows [39, 41]:

- Step 1: Generate initial the population of fireflies placed at random positions within the n-dimensional search space.
- Step 2: Initialize the parameters, such as the light absorption coefficient (γ).

- Step 3: Define the light intensity (I_i) of each firefly (x_i) as the value of the cost function, $f(x_i)$.
- Step 4: For each firefly (x_i), compare its light intensity with the light intensity of every other firefly (e.g. x_j).
- Step 5: If ($I_j > I_i$), then move firefly x_i towards x_j in n-dimensions.
- Step 6: Calculate the new values of the cost function for each firefly and update the light intensity.
- Step 7: Rank the fireflies and determine the current best.
- Step 8: Repeat Steps 3–7 until the termination criteria is satisfied.

Data Mining using FA. The authors of [42] used FA to deal with clustering problem. A set of well-known and well-used benchmark data set have been used to test the performance. Compared with other algorithms (i.e., artificial bee colony and PSO), the simulation results showed that FA can be efficiently used for clustering.

3.4   Fireworks Optimization Algorithm (FOA)

In this section, we will introduce an emerging CI algorithm that is derived from the explosion process of fireworks, an explosive devices invented by our clever ancestor, which can produce striking display of light and sound [43].

Fireworks optimization algorithm (FOA) was recently proposed in [44]. The basic idea was when we need to find a point x_j satisfying $f(x_i) = y$, a set of fireworks will be continuously fired in the potential search space until an agent (i.e., a spark in fireworks context) gets to or reasonably close to the candidate point x_j. Based on this understanding, to implement FOA, the following steps need to be performed [44–46]:

- Step 1: Fireworks explosion process designing. Since the number of sparks and their coverage in the sky determines whether an explosion is good or not, Tan and Zhu [44] first defined the number of sparks created by each firework x_j as follows:

$$s_i = m \cdot \frac{y_{max} - f(x_i) + \xi}{\sum_{i=1}^{n} [y_{max} - f(x_i)] + \xi}. \tag{16}$$

where m is a parameter used to control the total number of sparks created by the n fireworks, $y_{max} = \max(f(x_i))$ (for $i = 1, 2, \ldots, n$) stands for the maximum value of the objective function among the y_{max} fireworks, and ξ represents a small constant which is used to avoid zero-division-error.

Meanwhile, in order to get rid of the overwhelming effects of the splendid fireworks, bounds s_i are also defined as follows [44]:

$$\hat{s}_i = \begin{cases} round(a \cdot m) & \text{if } s_i < am \\ round(b \cdot m) & \text{if } s_i > bm, \ a < b < 1. \\ round(s_i) & \text{otherwise} \end{cases} \tag{17}$$

where a and b are constant parameters.

Next, Tan and Zhu [44] also designed the explosion amplitude as follows:

$$A_i = \hat{A} \cdot \frac{f(x_i) - y_{\min} + \xi}{\sum_{i=1}^{n} [f(x_i) - y_{\min}] + \xi}. \tag{18}$$

where $\hat{A}$ represents the maximum amplitude of an explosion, and $y_{\min} = \min(f(x_i))$ (for $i = 1, 2, \ldots, n$) denotes the minimum value of the objective function among the n fireworks.

Finally, the directions of the generated sparks are computed as follows [44]:

$$z = round(d \cdot rand(0, 1)). \tag{19}$$

where d denotes the dimensionality of the location x, and $rand(0, 1)$ represents an uniformly distributed number within [0, 1].

- Step 2: In order to obtain a good implementation of FOA, the locations of where we want the fireworks to be fired need to be chosen properly. According to [44], the general distance between a location x and other locations can be expressed as follows:

$$R(x_i) = \sum_{j \in K} d(x_i, x_j) = \sum_{j \in K} \|x_i - x_j\|. \tag{20}$$

where K denotes a group of current locations of all fireworks and sparks.

The selection probability of a location x_i is then defined as follows [44]:

$$p(x_i) = \frac{R(x_i)}{\sum_{j \in K} R(x_j)}. \tag{21}$$

Data Mining using FOA. To validate the performance of the proposed FOA, 9 benchmark test functions were chosen by Tan and Zhu [44] and the comparisons were conducted among the standard PSO, and the clonal PSO. The experiment results indicated that the FOA clearly outperforms the other algorithms in both optimization accuracy and convergence speed. In addition, the authors of [46] used three sampling data methods to approximate fitness landscape in order to accelerating FOA.

3.5 Shark-Search Algorithm (SSA)

Shark-search algorithm (SSA) was originally proposed in [47] for enhancing the Web browsing and search engine performance [48–50]. There are several variants and applications of SSA can be found in the literature [51–53]. The SSA is built on its predecessor called "fish-search" algorithm and the key principles underlying them are the following:

The algorithm first takes an input as a seed URL (standing for uniform resource locator) and search query. Then it dynamically sets up a priority sequence for the next URLs to be explored. At each step, the first node is popped from the list and attended. As each file's text becomes available, it will be analyzed by a scoring component and then evaluated for its relevance or irrelevance to the search query.

Putting it simply, in SSA, when relevant information (standing for food) is discovered, searching agents (i.e., fish) reproduce and keep looking for food. They will die when the food is in absent condition or encountering polluted water (poor bandwidth situation). The original fish-search algorithm suffers from the following limitations [47]:

- First, the relevance score is assigned in a discrete way.
- Second, the differentiation degree of the priority of the pages in the list is very low.
- Third, the number of addressed children is reduced by arbitrary using the width parameter.

Bearing this in mind, there are several improvements provided in SSA [47]:

- Improvement 1: A similarity engine is introduced to rank the document relevance degree.
- Improvement 2: Refining the computation of the potential score of the children by taking the following two factors into account. First, propagating ancestral relevance scores deeper down the hierarchical structure. Second, making use of the meta-information carried by the links to the files.

Data Mining using SSA. To evaluate the efficacy of SSA, a measure called "getting as many relevant documents with the shorted delays" was proposed in [47]. By testing SSA on four case studies, the significant improvements were experimentally verified which offer SSA an advantage to replace original fish-search algorithm in dealing with dynamic and precise searches within the limited time range.

3.6 FlockbyLeader Algorithm

The FlockbyLeader algorithm was proposed by Bellaachia and Bari [54] in which the recently discovered leadership dynamic mechanisms in pigeon flocks are incorporated in the normal flocking model (i.e., Craig Reynolds' Model [55]). In every iteration, the algorithm starts by finding flock leaders. The main steps are illustrated as follows [54]:

- Calculating fitness value of each flock leader (L_i) according to the objective function (i.e., $d_{max}^{L_i}$). It will be defined as follows [54]:

$$d_{max}^{L_i} = \max_{o \in kNB_t(x_i)} \{\rho(x_i, o)\}. \tag{22}$$

where $kNB_t(x_i)$ is the k-neighbourhood of x_i at iteration t, $d_{max}^{L_i}$ as radius associated with leader L_i at iteration t, x_i is a node in the feature graph, and $\rho(x_i, o)$ is the given distance function between objects x_i and o.

- Ranking the LeaderAgent (A_i). This procedure is defined as follows, respectively [54]:

$$Rank_t(A_i) = \text{Log}\left(\frac{|N_{i,t}|}{|N_t|} \cdot 10\right) \cdot ARF_t(A_i). \tag{23}$$

$$ARF_t(A_i) = \frac{|DR\_kNB_t(x_i)|}{|DR\_kNB_t(x_i)| + |D\_kNB_t(x_i)|}. \tag{24}$$

$$\begin{cases} \text{if } ARF_t(A_i) \geq 0.5, & \text{then } x_i \text{ is a flockleader} \\ \text{if } ARF_t(A_i) < 0.5, & \text{then } x_i \text{ is a follower} \end{cases} \tag{25}$$

where $DR\_kNB_t(x_i)$ represents the dynamic reverse k-neighbourhood of x_i at iteration t, $ARF_t(A_i)$ is the dynamic agent role factor of the agent A_i at iteration t, $|N_{i,t}|$ is the number of the neighbours A_i at iteration t, and $|N_t|$ is the number of unvisited nodes at iteration t.

- Performing the flocking behaviour.
- Updating the FindFlockLeaders (G_f).

Data Mining using FlockbyLeader Algorithm. To test the efficiency of the FlockbyLeader algorithm, two large datasets that one is consists of 100 news articles collected from cyberspace, and the other one is the Iris Plant Dataset were adopted by Bellaachia and Bari [54]. Compared with other CI algorithms, the proposed algorithm is significant in improving the results.

3.7 Flocking-Based Algorithm (FBA)

Flocking-based algorithm (FBA) was originally proposed in [56–58]. The basic flocking model is composed of three simple steering rules (see below) that need to be executed at each instance over time, for each individual agent.

- Rule 1: Separation. Steering to avoid collision with other boids nearby.
- Rule 2: Alignment. Steering toward the average heading and speed of the neighboring flock mates.
- Rule 3: Cohesion. Steering to the average position of the neighboring flock mates.

In the proposed algorithm, a fourth rule is added as below:

- Rule 4: Feature similarity and dissimilarity rule. Steering the motion of the boids with the similarity among targeted objects.

All these four rules can be formally express by the following equations [56]:

- The function of separation rule is to act as an active boid trying to pull away before crashing into each other. The mathematical implementation of this rule is thus can be described as follows [56]:

$$d(P_x, P_b) \leq d_2 \Rightarrow \vec{v}_{sr} = \sum_{x}^{n} \frac{\vec{v}_x + \vec{v}_b}{d(P_x, P_b)}. \tag{26}$$

where v_{sr} is velocity driven by Rule 1, d_2 is the distance pre-defined, v_b and v_x are the velocities of boids B and X.
- The function of alignment rule is to act as the active boid trying to align its velocity vector with the average velocity vector of the flock in its local neighbourhood. The degree of locality of this rule is determined by the sensor range of the active flock boid. This rule can be presented in a mathematical way through the following equation [56]:

$$d(P_x, P_b) \leq d_1 \cap d(P_x, P_b) \geq d_2 \Rightarrow \vec{v}_{ar} = \frac{1}{n} \sum_{x}^{n} \vec{v}_x. \tag{27}$$

where v_{cr} is velocity driven by Rule 3, d_1 and d_2 are pre-defined distance, and $(P_x - P_b)$ calculates a directional vector point.
- The flock boid tries to stay with the other boids that share the similar features with it. The strength of the attracting force is proportional to the distance (between the boids) and the similarity (between the boids' feature values) which can be expressed as follows [56]:

$$v_{ds} = \sum_{x}^{n} (S(B, X) \times d(P_x, P_b)). \tag{28}$$

where v_{ds} is the velocity driven by feature similarity, $S(B, X)$ is the similarity value between the features of boids B and X.
- The flock boid attempts to stay away from other boids with dissimilar features. The strength of the repulsion force is inversely proportional to the distance (between the boids) and the similarity value (between the boids' features) which are defined as follows [56]:

$$v_{dd} = \sum_{x}^{n} \frac{1}{S(B, X) \times d(P_x, P_b)}. \tag{29}$$

where v_{dd} is the velocity driven by feature dissimilarity. To get comprehensive flocking behavior, the actions of all the rules are weighted and summed to obtain a net velocity vector required for the active flock boid using the following equation [56]:

$$v = w_{sr}v_{sr} + w_{ar}v_{ar} + w_{cr}v_{cr} + w_{ds}v_{ds} + w_{dd}v_{dd}. \tag{30}$$

where v is the boid's velocity in the virtual space, and $w_{sr}, w_{ar}, w_{cr}, w_{ds}, w_{dd}$ are pre-defined weight values.

Data Mining using FBA. Document clustering is an essential operation used in unsupervised document organization, automatic topic extraction, and information retrieval. It provides a structure for organizing large bodies of data (in text form) for efficient browsing and searching. The authors of [56] utilized FBA for document clustering analysis. A synthetic data set and a real document collection (including 100 news articles collected from the Internet) were used in their study. In the synthesis data set, four data types were included with each containing 200 2-dimensional (x, y) data objects. Parameters x and y are distributed according to Normal distribution $N(\mu, \sigma)$; while for the real document collection data set, 100 news articles collected from the Internet at different time stages were categorized by human experts and manually clustered into 12 categories such as Airline safety, Iran Nuclear, Storm Irene, Volcano, and Amphetamine. In order to reduce the impact of the length variations of different documents, the authors of [56] further normalized each file vector to make it in unit length. Each term stands one dimension in the document vector space. The total number of terms in the 100 stripped test files is thus 4790 (i.e., 4790 dimensions). The experimental studies were carried out on the synthetic and the real document collection data sets, respectively, among FBA and other popular clustering algorithms such as ant clustering algorithm and K-means algorithm. The final testing results illustrated that the FBA can have better performance with fewer iterations in comparison with the K-means and ant clustering algorithm. In the meantime, the clustering results generated by FBA were easy to be visualized and recognized even by an untrained human user.

3.8 Fruit Fly Optimization Algorithm (FFOA)

Fruit fly optimization algorithm (FFOA) was originally proposed in [59, 60] that is based on the food foraging behaviour of fruit fly. Generally, the procedures of FFOA are described as follows [59]:

- Initialization phase: The fruit flies are randomly distributed in the search space (*InitX_axis* and *InitY_axis*) via the following equations, respectively [59]:

$$X_i = X\_axis + RandomValue. \tag{31}$$

$$Y_i = Y\_axis + RandomValue. \tag{32}$$

where the term "*RandomValue*" is a random vector that were sampled from a uniform distribution.

- Path construction phase: The distance and smell concentration value of each fruit fly can be defined as follows, respectively [59]:

$$Dist_i = \sqrt{X_i^2 + Y_i^2}. \tag{33}$$

$$S_i = \frac{1}{Dist_i}. \tag{34}$$

where $Dist_i$ is the distance between the ith individual and the food location, and S_i is the smell concentration judgment value which is the reciprocal of distance.
- Fitness function calculation phase. It can be defined as follows, respectively [59]:

$$Smell_i = Function(S_i). \tag{35}$$

$$[bestSmell, bestIndex] = \max(Smell_i). \tag{36}$$

where $Smell_i$ is the smell concentration of the individual fruit fly, $bestSmell$ and $bestIndex$ represent the largest elements and its indices along different dimensions of smell vectors, and $\max(Smell_i)$ is the maximal smell concentration among the fruit flies.
- Movement phase: The fruit fly keeps the best smell concentration value and will use vision to fly towards that location via the following equations, respectively [59]:

$$Smellbest = bestSmell. \tag{37}$$

$$X\_axis = X(bestIndex). \tag{38}$$

$$Y\_axis = Y(bestIndex). \tag{39}$$

Overall, taking into account the key phases described above, the steps of implementing FFOA can be summarized as follows [59]:

- Step 1: Initialize the optimization problem and algorithm parameters.
- Step 2: Repeat till stopping criteria met. First, randomly select a location via distance and smell concentration judgment value. Second, calculate its fitness function $Function(S_i)$. Third, find out the fruit fly with maximal smell concentration among the fruit fly swarm. Fourth, rank the solutions and move to the best solution.
- Step 3: Post process and visualize results.

Data Mining using FFOA. In order to show how the FFOA performs, two functions (i.e., one minimum and one maximum) are tested in [59]. In addition, the authors used FFOA to deal with the financial distress data of Taiwan's enterprise, computational results showed that FFOA has a very good classification and prediction capability.

3.9 Glowworm Swarm Optimization (GSO) Algorithm

Also inspired by luminous insect, the glowworm swarm optimization (GSO) algorithm was originally proposed by Krishnanand and Ghose [61] to deal with multimodal problems. Just like ants, elephants, mice, and snakes, glowworms also use some chemical substances, called luciferin, as signals for indirect communication. By sensing luciferin, glowworms can be attracted by strongest luciferin concentrations. In this way, the final optimization results can be found.

Typically, each iteration of the GSO algorithm consists of two phases, namely, a luciferin-update phase and a movement phase. In addition, for GSO, there is a dynamic decision range update rule that is used to adjust the glowworms' adaptive neighbourhoods. The details are listed as below [62]:

- Luciferin-update phase: It is the process by which the luciferin quantities are modified. The quantities value can either increase, as glowworms deposit luciferin on the current position, or decrease, due to luciferin decay. The luciferin update rule is given as follows [62]:

$$l_i(t+1) = (1 - \rho) \cdot l_i(t) + \gamma \cdot J \cdot [x_i \cdot (t+1)]. \tag{40}$$

  where $l_i(t)$ denotes the luciferin level associated with the glowworm i at time t, ρ is the luciferin decay constant $(0 < \rho < 1)$, γ is the luciferin enhancement constant, and $J(x_i(t))$ stands for the value of the objective function at glowworm i's location at time t.

- Movement phase: During this phase, glowworm i chooses the next position j to move to using a bias (i.e., probabilistic decision rule) toward good-quality individual which has higher luciferin value than its own. In addition, based on their relative luciferin levels and availability of local information, the swarm of glowworms can be partitioned into subgroups that converge on multiple optima of a given multimodal function. The probability of moving toward a neighbour is given as follows [62]:

$$p_{ij}(t) = \frac{l_j(t) - l_i(t)}{\sum_{k \in N_i(t)} [l_k(t) - l_i(t)]}. \tag{41}$$

where $j \in N_i(t)$, $N_i(t) = \{j : d_{ij}(t) < r_d^i(t); l_i(t) < l_{js}(t)\}$ is the set of neighbours of glowworm i at time t, $d_{ij}(t)$ denotes the Euclidean distance between glowworms i and j at time t, and $r_d^i(t)$ stands for the variable neighbourhood range associated with glowworm i at time t.

Based on the above equation, the discrete-time model of the glowworm movements can be stated as follows [62]:

$$x_i(t+1) = x_i(t) + s\left[\frac{x_j(t) - x_i(t)}{\|x_j(t) - x_i(t)\|}\right]. \tag{42}$$

where $x_i(t) \in \mathbf{R}^m$ is the location of glowworm i at time t in the m–dimensional real space, $\|\cdot\|$ denotes the Euclidean norm operator, and $s(> 0)$ is the step size.

- Neighbourhood range update rule: In addition to the luciferin value update rule that is illustrated in the movement phase, in GSO the glowworms use a radial range (i.e., $(0 < r_d^i \le r_s)$) update rule to explore an adaptive neighbourhood (i.e., to detect the presence of multiple peaks in a multimodal function landscape). Let r_0 be the initial neighbourhood range of each glowworm (i.e., $r_d^i(0) = r_0 \,\forall i$), then the updating rule is given as follows [62]:

$$r_d^i(t+1) = \min\{r_s, \ \max\{0, \ r_d^i(t) + \beta(n_t - |N_i(t)|)\}\}. \tag{43}$$

where β is a constant parameter, and $n_t \in N$ is a parameter used to control the number of neighbours.

Furthermore, in order to escape the dead-lock situation (i.e., all the glowworms converge to suboptimal solutions), Krishnanand and Ghose [62] employed a local search mechanism. The working principle is described as follows: during the movement phase, each glowworm moves a distance of step size (s) toward a neighbour. Hence, when $d_{ij}(t) < s$, glowworm i leapfrogs over the position of a neighbour j and becomes a leader to j. In the next iteration, glowworm i remains stationary and j overtakes the position of glowworm i, thus regaining its leadership. In this way, the GSO algorithm converges to a state in which all the glowworms construct the optimal solution over and over again.

Typically, by taking into account the basic rules described above, the steps of implementing the GSO algorithm can be summarized as follows [62]:

- Step 1: Initialize the parameters.
- Step 2: Initiation population of N candidate solution is randomly generated all over the search space.
- Step 3: The fitness function value corresponding to each candidate solution is calculated.
- Step 4: Perform the iteration procedures that include luciferin update phase, movement phase, and decision range update phase.
- Step 5: Check if maximum iteration is reached, go to step 3 for new beginning. If a specified termination criteria is satisfied, stop and return the best solution.

Data Mining using GSO. Based on GSO, the authors of [63] proposed two cluster analysis methods deal with data mining. Experimental results showed that the proposed algorithms have much potential.

3.10   Glowworm Swarm Optimization (GSO) Algorithm

Harmony search (HS) algorithm was originally proposed in Geem, Kim [64]. With the underlying fundamental of natural musical performance processes in which the musicians improvise their instruments' pitch by searching for the pleasing harmony (a perfect state), HS find the solutions through the determination of an objective function (i.e., the audience's aesthetics) in which a set of values (i.e., the musicians) assigned to each decision variable (i.e., the musical instrument's pitch). In general, the HS algorithm has three main operations: harmony memory (HM) consideration, pitch adjustment, and randomization [64]. The HS algorithm is performed in several steps, outlined below [64]:

- Preparation of harmony memory: The main building block of HS is the usage of HM, because multiple randomized solution vectors are stored in HM as follows [65]:

$$
\mathrm{HM} = \begin{bmatrix}
D_1^1 & D_2^1 & \cdots & D_n^1 & f(\mathbf{D}^1) \\
D_1^2 & D_2^2 & \cdots & D_n^2 & f(\mathbf{D}^2) \\
\vdots & \vdots & \cdots & \vdots & \vdots \\
D_1^{HMS} & D_2^{HMS} & \cdots & D_n^{HMS} & f(\mathbf{D}^{HMS})
\end{bmatrix}.
\tag{44}
$$

  where D_i^j is the ith decision variable in the jth solution vector that has one discrete value out of a candidate set $\{D_i(1), D_i(2), \ldots, D_i(k), \ldots, D_i(K_i)\}$, $f(\mathbf{D}^j)$ is the objective function value for the jth solution vector, and *HMS* is the harmony memory size (i.e., the number of multiple vectors stored in the HM).
- Improvisation of new harmony: A new harmony vector $D_i^{new} = (D_1^{new}, D_2^{new}, \ldots, D_n^{new})$ is improvised by the following three rules [65]:

  1. Random selection: Based on this rule, one value is chosen out of the candidate set as follows [65]:

$$
D_i^{new} \leftarrow D_i(k), \ D_i(k) \in \{D_i(1), D_i(2), \ldots, D_i(K_i)\}.
\tag{45}
$$

  2. HM consideration: In memory consideration, one value is chosen out of the HM set with a probability of harmony memory consideration rate (HMCR) as follows [65]:

$$
D_i^{new} \leftarrow D_i(l), \ D_i(l) \in \{D_i^1, D_i^2, \ldots, D_i^{HMS}\}.
\tag{46}
$$

  3. Pitch adjustment: According to this rule, the obtained value as in Eq. 46 is further changed into neighbouring values, with a probability of pitch adjusting rate (PAR) as follows [65]:

$$D_i^{new} \leftarrow D_i(l \pm 1), \ D_i(l) \in \{D_i^1, D_i^2, \cdots, D_i^{HMS}\}. \tag{47}$$

Overall, these three rules are the core terms of the stochastic derivative of HS and can be summarized as follows [65]:

$$\left.\frac{\partial f}{\partial D_i}\right|_{D_i = D_i(l)} = \frac{1}{K_i} \cdot (1 - HMCR)$$

$$+ \frac{n(D_i(l))}{HMS} \cdot HMCR \cdot (1 - PAR)$$

$$+ \frac{n(D_i(l \pm 1))}{HMS} \cdot HMCR \cdot PAR \tag{48}$$

where $\frac{1}{K_i} \cdot (1 - HMCR)$ denotes for the rate to choose a value $D_i(l)$ for the decision variable D_i by random selection, $\frac{n(D_i(l))}{HMS} \cdot HMCR \cdot (1 - PAR)$ chooses the rate by HM consideration, and $\frac{n(D_i(l \pm 1))}{HMS} \cdot HMCR \cdot PAR$ chooses the rate by pitch adjustment.

- Update of HM: Once the new vector $D_i^{new} = \left(D_1^{new}, D_2^{new}, \cdots, D_n^{new}\right)$ is completely generated, it will be compared with the other vectors that stored in HM. If it is better than the worst vector in HM with respect to the objective function, it will be updated (i.e., the new harmony is included in the HM and the existing worst harmony is excluded from the HM).

In summary, the general optimization procedure of the HS algorithm can be given as follows [64, 66]:

- Step 1: Initialize the optimization problem and algorithm parameters.
- Step 2: Initialization of HM.
- Step 3: Improvise a new harmony from the HM.
- Step 4: Update the HM.
- Step 5: Repeat Steps 3 and 4 until the termination criterion is satisfied.

Data Mining using HS. In terms of data mining, the authors of [67] proposed a novel harmony K-means algorithm which based on HS for document clustering. Compared with other optimization methods, the proposed algorithm is capable of convergence to the best known optimum faster than others. In a similar vein, the authors of [68] and [69] hybridized K-means and HS for clustering web documents. Experimental results revealed that the proposed algorithm can find better clusters when compared to similar methods. To deal with classification accuracy, studies, such as [70] and [71], made a preliminary attempt. Also, HS is used to solving the feature selection problem, such as [72] and [73].

3.11 Human Group Formation (HGF) Algorithm

Human group formation (HGF) algorithm was recently proposed in [74]. The key concept of this algorithm is about the behaviour of in-group members that try to unite with their own group as much as possible, and at the same time maintain social distance from the out-group members. To implement the HGF algorithm, the following steps need to be performed [74]:

- Step 1: Cluster centres representation refers to the number of classes, number of available input patterns, and number, type, and scale of the features available to the clustering algorithm. At first, there are a total of Q clusters, which is equal to the number of target output classes.
- Step 2: Accuracy selection is usually measured by a distance function defined on pairs of patterns as follows [74]:

$$\text{Accuracy} = \frac{\sum_{i=1}^{p} A_i}{P}$$

$$A_i = \begin{cases} 1, & \text{if } J \in Y_i \\ 0, & \text{otherwise} \end{cases} \tag{49}$$

$$J = \arg_j \min(d_j(X_i)), d_j(X_i) = \left\| X_i - z_j \right\|$$

where P denotes the total number of patterns in the training data set; J represents the index of a cluster whose reference pattern is the closest match to the incoming input pattern X_i; Y_i stands for the target output of the ith input pattern; z_j refers to the centre of the jth cluster; and $d_j(X_i)$ states the Euclidean distance between the input pattern X_i and the centre of the jth cluster.

- Step 3: The grouping/formation step can be performed in a way that in-group member try to unite with their own group and maintain social distance from the non-members as much as possible, update the centre value of each cluster (Z_j) as follows [74]:

$$Z_{jk}^{new} = Z_{jk}^{old} + \Delta Z_{jk}$$

$$\Delta Z_{jk} = \sum_{m \in q} \eta_{jm} \beta_j \delta_{jm} (Z_{mk} - Z_{jk}) - \sum_{n \notin q} \eta_{jn} \beta_j \delta_{jn} (Z_{nk} - Z_{jk}) \cdot \tag{50}$$

where k ($k = 1, 2, 3, \ldots, k$) is the number of features in the input pattern; q is the class to which the jth cluster belongs; $\eta_{jm} = e^{-\left[(Z_{jk} - Z_{mk})/\sigma\right]^2}$ and $\eta_{jn} = e^{-\left[(Z_{jk} - Z_{nk})/\sigma\right]^2}$ have values between 0 and 1 which determine the influence of mth and nth clusters on the jth cluster. In general, the further apart mth and nth clusters are from the jth cluster, the lower the values of η_{jm} and η_{jn}; β_j is the velocity of the jth cluster with respect to its own ability to move in the search space; and δ_{jm} is the parameter to prevent clusters of the same class from being too close to one another and normally with respect to two factors: (1) the distance between the jth cluster and the mth cluster, and (2) the territorial boundary of the clusters (T). If the distance between the jth cluster and the mth cluster is less than T, the value of δ_{jm} will be decreased by a predefined amount. After each centre is updated, if the accuracy is higher, save this new center value and then continue updating the next cluster centre; if it is lower, discard the new center value and return to the previous centre; and if it does not change, save the new center value and decrease the value of β_j by a predefined amount.

- Step 4: Cluster validity analysis is the assessment of clustering procedure's output. The cluster which satisfies the following equation will be deleted [74]:

$$-\frac{1}{2\log_2\left(\frac{n_j}{p}\right)}\left(\frac{n_j^q}{n_j}\right)\left(\frac{\sum_{\forall X_i^j \in q}\left\|X_i^j - z_j\right\|}{n_j}\right) < \rho. \tag{51}$$

where n_j is the number of input patterns in the jth cluster; n_j^q is the number of input patterns in the jth cluster whose target outputs (Y) are q; X_i^j is the ith input pattern in the jth cluster; and ρ is the vigilance parameter.

- Step 5: Recalculating the accuracy of the model according to Eq. 49 [74].
- Step 6: For each remaining cluster, if the distance between the new centre updated in step 3 and the previous centre is less than 0.0001 ($\left\|Z_{jk}^{new} - Z_{jk}^{old}\right\| < 0.0001$), randomly pick k small numbers between -0.1 and 0.1, and then add them to the centre value of the cluster. The purpose of this step is to prevent the premature convergence of the proposed algorithm to sub-optimal solutions.
- Step 7: Terminating process is to check the end condition, if it is satisfied, stop the loop; if not, examine the following conditions: (1) if the accuracy of the model improves over the previous iteration, randomly select one input pattern from the training data set of each target output class that still has error. Then go to step 2; and (2) if the accuracy does not improve, randomly select the input patterns, a number equal to the number of clusters deleted in step 4, from the training data set of each target output class. Then go to step 2.

Data Mining using HGF. One instinctive ability of HGF algorithm is classification. The authors of [74] used HGF to deal with four artificial and twelve real-life data sets, such as iris data, wine recognition data, and Haberman's survival data. Compared with other well-known classification methods, experimental results showed that HGF performs very well in all problems in terms of the classification accuracy and the size of the model.

3.12 Photosynthetic Algorithm (PA)

Motivated by the principle of Benson-Calvin cycle Phase-1 and the reaction that happens in the chloroplast subcellular compartment for photorespiration, photosynthetic algorithm (PA) was originally proposed in [75]. To perform the PA, the following calculation processes need to be followed [75]:

- First, randomly generating the intensity of light.
- Second, evaluating the fixation rate of CO_2 via the following equation (also refer to as the stimulation function in the PA algorithm) based on the light intensity [75]. This is a unique characteristic of the PA algorithm. Such stimulation often happens as a result of randomly changed light intensity which in turn adjusts the influential degree on the elements of RuBP (i.e., ribulose-1, 5-bishosphate [76]) by photorespiration.

$$C = \frac{V_{\max}}{1 + A/L}.\tag{52}$$

where the CO_2 fixation rate is denoted by C, $V_{\max}$ represents the maximum CO_2 fixation rate, A stands for the affinity of CO_2, and L is used to express the light intensity.

- Third, based on the fixation rate obtained from the stage above, one of two cycles, either Benson-Calvin or photorespiration will be selected at this stage. For both cycles, Murase [75] utilized 16-bit strings which shuffles based on carbon molecules recombination rule in photosynthetic pathways.
- Then after certain rounds of iterations, an amount of GAPs, i.e., glyceraldehyde-3-phosphate [77], are generated for representing intermediate knowledge strings in the PA algorithm. Each GAP is composed of 16 bits. The fitness of these GAPs will be evaluated at this stage. The best fit GAP will remain as a DHAP (i.e., di-hydroxyacetone phosphate [76]) which is referred to as the current estimated value.

Taking into account the fundamental process described above, the steps of implementing PA can be summarized as follows [75, 78, 79]:

- Step 1: Initializing the following problem parameters such as $f(x)$ (the object function), x_i (the decision variable), N (the number of decision variables), and the boundary of constraints.
- Step 2: Initializing the following problem parameters such as DHAPs, and CO_2 fixation parameters (e.g., affinity A, maximum fixation rate $V_{\max}$, and light intensity L).
- Step 3: Calculating CO_2 concentration, determining O_2/CO_2 concentration ration, and setting Benson-Calvin/photorespiration frequency ratio.
- Step 4: Evaluating if the stopping criteria are met. If yes, the algorithm stops; otherwise, go to the next step.
- Step 5: Depending the fixation rate of CO_2, the 16-bit strings are shuffled in either Benson-Calvin or photorespiration cycle.
- Step 6: Comparing the fitness value where the poor results will be removed and the desired DHAP strings and results will be remained.
- Step 7: Updating the light intensity and the next round of iteration of the PA algorithm starts.

Data Mining using PA. In order to verify the proposed PA, the finite element inverse analysis problem was employed in [75]. The prediction of the elastic moduli of the finite element model via PA was quite satisfied. The overall performance demonstrated by this preliminary application make PA a very attractive optimization algorithm. In data mining domain, the author of [78] used PA to solve multiple sequence alignment and association rules mining problem within biomedical data. Computational results showed that PA is capable of finding an effective global optima.

3.13 Stem Cells Optimization Algorithm (SCOA)

Stem cells optimization algorithm (SCOA) was originally proposed in [80, 81]. To perform the SCOA algorithm, the following procedure needs to be followed [80]:

- First, dividing the problem space into sections. The process can be accomplished totally in a random manner.
- Second, generating the initial population randomly and uniformly distributed in the whole search space of the target problem. At this stage, similar to most optimization algorithms, a variable matrix needs to be established for the purpose of obtaining a feedback with respect to problem variables. In SCOA, the key stem cells features are used to form the initial variable matrix. Such features may include liver cells, intestinal cells, blood cells, neurons, heart muscle cells, pancreatic islets cells, etc. Basically, the initial matrix can be express as equation below [80]:

$$\text{Population} = \begin{bmatrix} X_1 \\ X_2 \\ \cdots \\ X_N \end{bmatrix}. \tag{53}$$

where $X_i = \text{Stem Cells} = [SC_1, SC_2, \ldots, SC_N]; i = 1, 2, \ldots, N.$

In SCOA, some initialized parameters are defined as follows: M represents the maximum of stem cells; P stands for population size $(10 < P \leq M)$; $C_{Optimum}$ indicates the best of stem cell in each iteration; χ denotes the penalty parameter which is used to stop the growth of stem cell; and sc^i is the ith stem cell in the population.

- Third, the cost of each stem cell is obtained a criterion function which is determined based on the nature of the target problem. In SCOA, two types of memory, namely, local- and global-memory, are defined for each cell in which the local-memory is used to store the cost of each stem cell, and the global-memory stores the best cost among all locally stored cost values.
- Then, a self-renewal process will be performed which involves only the best cells of each area. At this stage, the information of each area's best cells will be shared and the cell that possesses the best cost will thus be chosen. In SCOA,

such cell is designed to play a more important role than other cells. Briefly, the stem cells' self-renewal operation is computed through equation below [80]:

$$SC_{Optimum}(t+1) = \zeta SC_{Optimum}(t). \tag{54}$$

where the iteration number is denoted by t, $SC_{Optimum}$ represents the best stem cell found in each iteration, and ζ is a random number which falls within $[0, 1]$.

- Next, the above mentioned procedure will continue until the SCOA arrives at the goal of getting the best cell while keeping the value of cost function as low as possible. This is acquired via equation below [80]:

$$x_{ij}(t+1) = \mu_{ij} + \varphi\big(\mu_{ij}(t) - \mu_{kj}(t)\big). \tag{55}$$

where the ith stem cell position for the solution dimension j is represented by x_{ij}, the iteration number is denoted by t, two randomly selected stem cells for the solution dimension j are denoted by μ_{ij} and μ_{kj}, respectively, and $\varphi(\tau)$ (if $\mu_{ij}(t) - \mu_{kj}(t) = \tau$) generates a random variable falls within $[-1, 1]$.

- Finally, the best stem cell is selected when it has the most power relative to other cells. The comparative power can be computed via equation below [80]:

$$\varsigma = \frac{SC_{Optimum}}{\sum_{i=1}^{N} SC_i}. \tag{56}$$

where ς stands for stem cells' comparative power, $SC_{Optimum}$ denotes the stem cells selected in terms of cost, and N represents the final population size, i.e., when the best solution is obtained and the SCOA algorithm terminates.

Data Mining using SCOA. The authors of [80] employed SCOA for data clustering. A set of well-known datasets have been used to test the performance. Compared with other methods such as artificial bee colony, PSO, and ACO, experimental results showed that SCOA demonstrates superior clustering performance in terms of accuracy and high speed.

3.14 Wasp Swarm Optimization (WSO) Algorithm

Wasp swarm optimization (WSO) algorithm was originally proposed in [82] that is based on some behaviours found in wasp colony [83, 84]. The basic idea of WSO was to mimic a wasp colony behaviour, in particular according to the importance of individual wasp to the whole colony, assigning the resources to different wasp [82, 85]. Therefore in WSO algorithm, resources will be allocated to individual candidate solutions and such allocation is completed in a randomly manner where the strength of each option controls its chosen probability. In [86], a tournament process was utilized to implement this stochastic selection process: The weakest option (for example a) challenges the second weakest option (for instance b) and the winning probability of a over b is determined through $p_{ab} = s_a^2 / (s_a^2 + s_b^2)$. The winner of this challenge (say a) will carry on to challenges the third weakest option (denoted by c), and wins with a probability of $p_{ac} = s_a^2 / (s_a^2 + s_c^2)$. The challenge will continue until the final winner is selected. In some situations, it is more convenient to allocate costs instead of strengths to the individual wasps, i.e., the lower the cost, the higher the strength of a wasp. In this case, the winning probability of wasp i over j can be defined via equation below [86]:

$$p_{ij} = \frac{s_j^2}{s_i^2 + s_j^2}, \ i, j = 1, \ldots, c. \tag{57}$$

Data Mining using WSO. In [87], the author used WSO to optimize the C-means clustering model. Compared with other algorithms such as ACO, WSO outperforms ACO in terms of robust and high speed.

4   Discussions

Like other essential techniques of data qualify such as capture, communicate, aggregate and store, it is increasingly the case that much of data mining techniques simply could not take place without innovation. A conceptual view of big data challenges is provided by [4]. In the proposed framework, there are three challenges related to big data management operations:

- Data-stream processing and actual computing procedures.
- Different semantics and domain knowledge for big data applications.
- Algorithms for creating, clustering, and analyzing big data.

Over the past decade, data mining based upon CI approaches has been a widely studied research topic, being capable of meeting the ever-increasing demand of reliable and intelligent data mining techniques.

In our view, the methods introduced in this study are all developed with their inspiring source more or less coming from our Mother Nature. Through the intelligence introduced via the different metaphor mechanisms, these clever algorithms can speculate the intrinsic patterns from data sets without or with limited prior knowledge of regularities that might existing in the data. Such speculation is normally realized through either learning or searching process.

5 Future Work

The present work has some limitations. Firstly, a widespread literature review of the applications of innovative CI presents a difficult task, because of the extensive background knowledge that is required during the process of collection, study, and classification of these publications. Although acknowledging a limited background knowledge, this research makes a brief overview of literature concerned with using biology-based innovative CI algorithms to deal with data mining. Secondly, there are also some algorithms that falls under the other categories such as physics-, chemistry-, mathematics-based innovative CI. However, the present study does not take them into account. This would be an immediately research direction that need to be considered in future study.

6 Conclusion

Ubiquitous robotics systems has grown very fast in recent years. This is a good news from many perspectives, in particular considering the huge challenges confronted with the rapidly aging population. However, this remarkable growth also comes with some critical technical challenges. Among others, a lack of knowledge on how to manage the constant flow of data is probably a daunting disadvantages. As a result, data mining techniques are gradually being introduced into ambient intelligent system as the key to better understanding user behaviour and as a layer of intelligence for their own manipulation.

Nowadays, compared to the standard data management, big data, which is generated by multiple velocity sources and volumes, poses a considerable threat to a successful ambient intelligent system's implementation. To effectively and efficiently analyse the ever-increasing massive amount of data, on one hand, we can enhance the capability of the existing data mining techniques; while on the other hand, developing more powerful versions of data mining algorithms is also becoming extremely important, in particular, when many of the traditional approaches are reaching the limit of their full potential.

By dedicating to the latter trend, this study presents an up-to-date overview regarding the development of innovative CI algorithms and their corresponding applications in data mining field. Through our investigation, although the preliminary practice of these approaches cannot yet reveal that the traditional data mining techniques are overall outperformed by their newly introduced counterparts, these recently joined CI family members do show various attractive and promising advantages in dealing with data mining. Therefore it is the author's hope that this survey will inspire other far better qualified researchers to bring these algorithms to their full potential for embracing the forthcoming ambient intelligence era.

References

1. Manyika, J., et al.: Big data: the next frontier for innovation, competition, and productivity. McKinsey Global Institute (2011)
2. Gobble, M.M.: Big data: the next big thing in innovation. Res. Technol. Manage. **56**(1), 64–66 (2013)
3. Katal, A., Wazid, M., Goudar, R.H.: Big data: issues, challenges, tools and good practices. In: 2013 Sixth International Conference on Contemporary Computing (IC3), pp. 404–409. IEEE (2013)
4. Wu, X., et al.: Data mining with big data. IEEE Trans. Knowl. Data Eng. **26**(1), 97–107 (2014)
5. Nugent, C.D., et al.: Managing sensor data in ambient assisted living. J. Comput. Sci. Eng. **5** (3), 237–245 (2011)
6. Yu, B., Sen, R., Jeong, D.H.: An integrated framework for managing sensor data uncertainty using cloud computing. Inf. Syst. **38**, 1252–1268 (2013)
7. Choudhary, A.K., Harding, J.A., Tiwari, M.K.: Data mining in manufacturing: a review based on the kind of knowledge. Int. J. Intell. Manuf. **20**, 501–521 (2009)
8. Ahmed, S.R.: Applications of data mining in retail business. Inf. Technol.: Coding Comput. **2**, 455–459 (2004)
9. Liao, S.-H., Chu, P.-H., Hsiao, P.-Y.: Data mining techniques and applications—a decade review from 2000 to 2011. Expert Syst. Appl. **39**, 11303–11311 (2012)
10. Liao, S.-H., Chen, Y.-J., Lin, Y.-T.: Mining customer knowledge to implement online shopping and home delivery for hypermarkets. Expert Syst. Appl. **38**, 3982–3991 (2011)
11. Ngai, E.W.T., Xiu, L., Chau, D.C.K.: Application of data mining techniques in customer relationship management: a literature review and classification. Expert Syst. Appl. **36**, 2592–2602 (2009)
12. Han, J., Kamber, M., Pei, J.: Data mining: concepts and techniques. 3rd ed. 2012, 225 Wyman Street, Waltham, MA 02451, Morgan Kaufmann, Elsevier Inc., USA. ISBN 978-0-12-381479-1 (2012)
13. Xing, B., Gao, W.-J.: Innovative computational intelligence: a rough guide to 134 clever algorithms. Springer International Publishing Switzerland, Cham, Heidelberg, New York, Dordrecht, London (2014). ISBN 978-3-319-03403-4
14. Elalfi, E., Haque, R., Elalami, M.E.: Extracting rules from trained neural network using GA for managing E-business. Appl. Soft. Comput. **4**, 65–77 (2004)
15. Xie, L., Mei, H.: The application of the ant colony decision rule algorithm on distributed data mining. Commun IIMA **7**(4), 85–94 (2007)
16. Sinha, A.N., Das, N., Sahoo, G.: Ant colony based hybrid optimization for data clustering. Kybernetes **36**(2), 175–191 (2007)
17. de Castro, L.N., Zuben, F.J.V.: aiNet: an artificial immune network for data analysis. In: Abbass, H.A., Sarker, R.A., Newton, C.S (ed.) Data mining: a heuristic approach, Idea Group Publishing, (2001)
18. Alatas, B., Akin, E.: Multi-objective rule mining using a chaotic particle swarm optimization algorithm. Knowl.-Based Syst. **22**, 455–460 (2009)
19. Mitra, S., Pal, S.K., Mitra, P.: Data mining in soft computing framework: a survey. IEEE Trans. Neural Netw. **13**(1), 3–14 (2002)
20. Romero, C. Ventura, S.: Educational data mining: a survey from 1995 to 2005. Expert Syst. Appl. **33**, 135–146 (2007)
21. Harding, J.A., et al.: Data mining in manufacturing: a review. J. Manuf. Sci. Eng. **128**, 969–976 (2006)
22. Glaser, R.: Biophysics: an introduction, 2nd edn. Springer, Berlin, Heidelberg (2012). ISBN 978-3-642-25211-2

23. Floreano, D., Mattiussi, C.: Bio-inspired artificial intelligence: theories, methods, and technologies. In: Arkin, R.C. (ed.) Intelligent Robotics and Autonomous Agents 2008, The MIT Press, Cambridge, Massachusetts. ISBN 978-0-262-06271-8 (2008)
24. Li, X-l: A new intelligent optimization method—artificial fish school algorithm (in Chinese with English abstract), in Institute of Systems Engineering. Zhejiang University, Hangzhou, P. R. China (2003)
25. Zhang, M., et al.: Evolving neural network classifiers and feature subset using artificial fish swarm. In: IEEE International Conference on Mechatronics and Automation 25–28 June, Luoyang, China, pp. 1598–1602. IEEE (2006)
26. Wang, C.-R., Zhou, C.-L., Ma J.-W.: An improved artificial fish-swarm algorithm and its application in feed-forward neural networks. In: Fourth International Conference on Machine Learning and Cybernetics, Guangzhou, China, 18–21 Aug, pp. 2890–2894. (2005)
27. Luo, Y., Zhang, J., Li, X.: The optimization of PID controller parameters based on artificial fish swarm algorithm. In: IEEE International Conference on Automation and Logistics, 18–21 Aug, Jinan, China, pp. 1058–1062. (2007)
28. Neshat, M., et al.: A review of artificial fish swarm optimization methods and applications. Int. J. Smart Sens. Intell. Syst. 5(1), 107–148 (2012)
29. Zhang, M., et al.: Mining classification rule with artificial fish swarm. In: 6th World Congress on Intelligent Control and Automation, 21–23 June, Dalian, China, pp. 5877–5881 (2006)
30. Wang, F., Xu, X., Zhang, J.: Application of artificial fish school and K-means clustering algorithms for stochastic GHP. In: Control and Decision Conference (CCDC), pp. 4280–4283. 2009
31. Neshat, M., et al.: A new hybrid algorithm based on artificial fishes swarm optimization and k-means for cluster analysis. Int. J. Comput. Sci. Issues 8(4), 251–259 (2011)
32. Sun, S., Zhang, J., Liu, H.: Key frame extraction based on artificial fish swarm algorithm and k-means. In: International Conference on Transportation, Mechanical, and Electrical Engineering (TMEE), 16–18 Dec, Changchun, China, pp. 1650–1653. IEEE (2011)
33. He, S., et al.: Fuzzy clustering with improved artificial fish swarm algorithm. In: International Joint Conference on Computational Sciences and Optimization (CSO), pp. 317–321. IEEE (2009)
34. Cheng, Y., Jiang, M., Yuan, D.: Novel clustering algorithms based on improved artificial fish swarm algorithm. In: Sixth International Conference on Fuzzy Systems and Knowledge Discovery (FSKD), pp. 141–145. IEEE (2009)
35. Zhu, W., et al.: Clustering algorithm based on fuzzy C-means and artificial fish swarm. Procedia Eng. 29, 3307–3311 (2012)
36. Su, M.-C., Su, S.-Y., Zhao, Y.-X.: A swarm-inspired projection algorithm. Pattern Recogn. 42, 2764–2786 (2009)
37. Yang, X.-S.: Firefly algorithm, stochastic test functions and design optimisation. Int. J. Bio-Inspired Comput. 2(2), 78–84 (2010)
38. Yang, X.-S.: Nature-inspired metaheuristic algorithms, 2nd edn. Luniver Press, UK (2008). ISBN 978-1-905986-28-6
39. Yang, X.-S.: Firefly algorithms for multimodal optimization. In: Watanabe, O., Zeugmann, T (ed.) SAGA 2009, LNCS 5792, pp. 169–178, Springer, Berlin, Heidelberg (2009)
40. Babu, B.G., Kannan, M.: Lightning bugs. Resonance 7(9), 49–55 (2002)
41. Jones, K.O., Boizanté, G.: Comparison of firefly algorithm optimisation, particle swarm optimisation and differential evolution. In: International Conference on Computer Systems and Technologies (CompSysTech), 16–17 June, Vienna, Austria, pp. 191–197. (2011)
42. Senthilnath, J., Omkar, S.N., Mani, V.: Clustering using firefly algorithm: performance study. Swarm and Evol. Comput. 1, 164–171 (2011)
43. Lancaster, R., et al.: Fireworks: principles and practice. Chemical Publishing Co., Inc., New York. ISBN 0-8206-0354-6 (1998)
44. Tan, Y. Zhu, Y.: Fireworks algorithm for optimization. In: Tan, Y., Shi, Y., Tan, K.C. (ed.) ICSI 2010, Part I, LNCS 6145, pp. 355–364, Springer, Berlin, Heidelberg (2010)

45. Janecek, A., Tan, Y.: Swarm intelligence for non-negative matrix factorization. Int. J. Swarm Intell. Res. **2**(4), 12–34 (2011)
46. Pei, Y., et al.: An empirical study on influence of approximation approaches on enhancing fireworks algorithm. In: IEEE International Conference on Systems, Man, and Cybernetics (IEEE SMC 2012), Seoul, Korea, 14–17 Oct, pp. 1322–1327. IEEE (2012)
47. Hersovici, M., et al.: The shark-search algorithm. An application: tailored Web site mapping. Comput. Netw. ISDN Syst. **30**, 317–326 (1998)
48. Hillis, K., Petit, M., Jarrett, K.: Google and the culture of search. Taylor & Francis, 711 Third Avenue, New York, NY: Routledge (2013). ISBN 978-0-415-88300-9
49. Cho, J., Garcia-Molina, H., Page, L.: Efficient crawling through URL ordering. Comput. Netw. ISDN Syst. **30**, 161–172 (1998)
50. Jarvis, J.: What whould Google do?. HarperCollins Publishers Ltd., 55 Avenue Road, Suite 2900, Toronto, ON, M5R, 3L2, Canada (2009). ISBN 978-0-06-176472-1
51. Luo, F.-f., Chen, G.-l., Guo W.-z.: An improved fish-search algorithm for information retrieval. In: IEEE International Conference on Natural Language Processing and Knowledge Engineering (IEEE NLP-KE), pp. 523–528. IEEE (2005)
52. Chen, Z., et al.: An improved shark-search algorithm based on multi-information. In: Fourth International Conference on Fuzzy Systems and Knowledge Discovery (FSKD), pp. 1–5. (2007)
53. Sun, T., et al.: Airplane route planning for plane-missile cooperation using improved fish-search algorithm. In: International Joint Conference on Artificial Intelligence (JCAI), pp. 853–856. (2009)
54. Bellaachia, A. Bari, A.: Flock by leader: a novel machine learning biologically inspired clustering algorithm. In: Tan, Y., Shi, Y., Ji, Z (ed.) ICSI 2012, Part I, LNCS 7332, pp. 117–126, Springer, Berlin, Heidelberg (2012)
55. Reynolds, C.W.: Flocks, herds, and schools: a distributed behavioral model. Comput. Graph. **21**(4), 25–34 (1987)
56. Cui, X., Gao, J., Potok, T.E.: A flocking based algorithm for document clustering analysis. J. Syst. Architect. **52**, 505–515 (2006)
57. Picarougne, F., et al.: A new approach of data clustering using a flock of agents. Evol. Comput. **15**(3), 345–367 (2007)
58. Luo, X., Li, S., Guan, X.: Flocking algorithm with multi-target tracking for multi-agent systems. Pattern Recogn. Lett. **31**, 800–805 (2010)
59. Pan, W.-T.: A new fruit fly optimization algorithm: taking the financial distress model as an example. Knowl.-Based Syst. **26**, 69–74 (2012)
60. Pan, W.-T.: Fruit fly optimization algorithm (in Chinese). Tsang Hai Book Publishing Co. ISBN 978-986-6184-70-3 (2011)
61. Krishnanand, K.N. Ghose, D:. Detection of multiple source locations using a glowworm metaphor with applications to collective robotics. In: IEEE Swarm Intelligence Symposium (SIS), IEEE, pp. 84–91 (2005)
62. Krishnanand, K.N., Ghose, D.: Glowworm swarm optimization for simultaneous capture of multiple local optima of multimodal functions. Swarm Intell. **3**, 87–124 (2009)
63. Huang, Z., Zhou, Y.: Using glowworm swarm optimization algorithm for clustering analysis. J. Convergence Inf. Technol. **6**(2), 78–85 (2011)
64. Geem, Z.W., Kim, J.H., Loganathan, G.V.: A new heuristic optimization algorithm: harmony search. Simulation **76**(2), 60–68 (2001)
65. Geem, Z.W.: Particle-swarm harmony search for water network design. Eng. Optim. **41**(4), 297–311 (2009)
66. Lee, K.S., Geem, Z.W.: A new meta-heuristic algorithm for continuous engineering optimization: harmony search theory and practice. Comput. Methods Appl. Mech. Eng. **194**, 3902–3933 (2005)
67. Mahdavi, M., Abolhassani, H.: Harmony K-means algorithm for document clusterin. Data Min. Knowl. Disc. **18**, 370–391 (2009)

68. Mahdavi, M., et al.: Novel meta-heuristic algorithms for clustering web documents. Appl. Math. Comput. **201**, 441–451 (2008)
69. Cobos, C., et al.: Web document clustering based on global-best harmony search, k-means, frequent term sets and Bayesian information crite. In: Proceedings of the IEEE World Congress on Computational Inteliigence (WCCI), 18–23 July 2010, CCIB, Barcelona, Spain, IEEE, pp. 4637–4644. 2010
70. Moeinzadeh, H., et al.: Combination of harmony search and linear discriminate analysis to improve classification. In: Proceedings of the Third Asia International Conference on Modelling & Simulation (AMS), IEEE, pp. 131–135 (2009)
71. Wang, X., Gao, X.-Z., Ovaska, S.J.: Fusion of clonal selection algorithm and harmony search method in optimization of fuzzy classification systems. Int. J. Bio-Inspired Comput. **1**, 80–88 (2009)
72. Venkatesh, S.K., Srikant, R., Madhu, R.M.: Feature selection & dominant feature selection for product reviews using meta-heuristic algorithms. In: Proceedings of the Compute'10, 22–23 Jan 2010, Bangalore, Karnataka, India, pp. 1–4. ACM (2010)
73. Ramos, C.C.O., et al.: A novel algorithm for feature selection using harmony search and its application for non-technical losses detection. Comput. Electr. Eng. **37**, 886–894 (2011)
74. Thammano, A., Moolwong, J.: A new computational intelligence technique based on human group formation. Expert Syst. Appl. **37**, 1628–1634 (2010)
75. Murase, H.: Finite element inverse analysis using a photosynthetic algorithm. Comput. Electron. Agric. **29**, 115–123 (2000)
76. Carpentier, R. (ed.) Photosynthesis research protocols. In: Walker, J.M., (ed.) 2nd edn. Methods in Molecular Biology, 2011, Springer Science+Business Media, LLC, ISBN 978-1-60671-924-6. Spring, New York, NY 10013, USA (2011)
77. Dubinsky, Z. (ed).: Photosynthesis. InTech, ISBN 978-953-51-1161-0: Janeza Trdine 9, 51000 Rijeka, Croatia (2013)
78. Alatas, B.: Photosynthetic algorithm approaches for bioinformatics. Expert Syst. Appl. **38**, 10541–10546 (2011)
79. Yang, X.-S.: Biology-derived algorithms in engineering optimization. In: Olarius, S., Zomaya, A. (ed.) Handbook of Bioinspired Algorithms and Applications, Chapter 32, pp. 585–596, CRC Press, LLC (2005)
80. Taherdangkoo, M., Yazdi, M., Bagheri, M.H.: A powerful and efficient evolutionary optimization algorithm based on stem cells algorithm for data clustering. Cent. Eur. J. Comput. Sci. **2**(1), 1–13 (2012)
81. Taherdangkoo, M., Yazdi, M., Bagheri, M.H: Stem cells optimization algorithm, in LNBI 6840, pp. 394–403. Springer, Berlin, Heidelberg (2011)
82. Theraulaz, G., et al.: Task differentiation in polistes wasps colonies: a model for self-organizing groups of robots. In: First International Conference on Simulation of Adaptive Behavior, MIT Press, pp. 346–355 (1991)
83. Karsai, I., Wenzel, J.W.: Organization and regulation of nest construction behavior in Metapolybia wasps. J. Insect Behav. **13**(1), 111–140 (2000)
84. Lucchetta, P., et al.: Foraging and associative learning of visual signals in a parasitic wasp. Anim. Cogn. **11**(3), 525–533 (2008)
85. Fan, H., Zhong, Y.: A rough set approach to feature selection based on wasp swarm optimization. J. Comput. Inf. Syst. **8**(3), 1037–1045 (2012)
86. Cicirello, V.A., Smith, S.F.: Wasp-like agents for distributed factory coordination. Auton. Agent. Multi-Agent Syst. **8**, 237–266 (2004)
87. Runkler, T.A.: Wasp swarm optimization of the c-means clustering model. Int. J. Intell. Syst. **23**, 269–285 (2008)

Author Biography

Bo Xing, DIng, is an Associate Professor at the Department of Computer Science, School of Mathematical and Computer Science, Faculty of Science and Agriculture, University of Limpopo, South Africa. He was a senior lecturer under the division of Center for Asset Integrity Management (C-AIM) at the Department of Mechanical and Aeronautic Engineering, Faculty of Engineering, Built Environment and Information Technology, University of Pretoria, South Africa. Dr. Xing earned his DIng degree (Doctorate in Engineering with a focus on soft computing and remanufacturing) in the early 2013 from the University of Johannesburg, South Africa. He also obtained his B.Sc. and M.Sc. degree both in Mechanical Engineering from the Tianjin University of Science and Technology, P.R. China, and the University of KwaZulu-Natal, South Africa, respectively. He was a scientific researcher at the Council for Scientific and Industrial Research (CSIR), South Africa. He has published more than 50 research papers in books, international journals, and international conference proceedings. His current research interests lie in applying various nature-inspired computational intelligence methodologies towards big data analysis, miniature robot design and analysis, advanced mechatronics system, and e-maintenance.

Ambient Stupidity

Jordi Vallverdú

Abstract AmI is clearly the next step into the evolution of the integration in daily life of computer technologies and AI products. Despite of the revolutionary nature of AmI, its analysis and media attention was big in the middle of 21st century decade, but decreasing since then. And this is a mistake, because the new social tendencies as well as the advancements in technological equipment and data procession techniques are allowing a next step into the advancement of AmI. Some general mistakes have been pointed and labeled as 'Ambient stupidity' and a broad philosophical analysis framework is suggested in order to prepare and improve the advent of *Big AmI*. At the same time, some critical remarks on the nature of human cognitive processes are introduced or reviewed. Finally, the request of new tools to deal with multivariate and dynamic sources of data is showed as a necessity of future researches.

Keywords Ambient intelligence · Philosophy · Epistemology · Stupidity · AI · Freedoom

1 The Basics of Ambient Intelligence

'Ambient Intelligence' (henceforth, AmI) is a concept that usually refers to a digital environment that proactively, but sensibly, supports people in their daily lives [1]. We will follow the [2] historical review in order to detail its geographical and conceptual emergence.

J. Vallverdú (✉)
Department of Philosophy, Universitat Autònoma de Barcelona,
Bellaterra (BCN), 08193 Barcelona, Catalonia, Spain
e-mail: Jordi.vallverdu@uab.cat
URL: https://uab.academia.edu/JordiVallverdu; http://orcid.org/0000-0001-9975-7780

© Springer International Publishing Switzerland 2016   173
K.K. Ravulakollu et al. (eds.), *Trends in Ambient Intelligent Systems*,
Studies in Computational Intelligence 633, DOI 10.1007/978-3-319-30184-6_7

1.1   The History of AmI

The first approach to AmI can be found in 1991, when Max Weiser, chief scientist at Xerox Palo Alto Research Center (PARC), published a paper in *Scientific American* in which he told about the evolution of computation and described a new era in which computers would be smaller than ever, interconnected and omnipresent. He called it "ubiquitous computing". Later, a European Philips' researcher, Emile Aarts, coined the term "Ambient Intelligence" [3], which shared most of the ideas of Weiser, but integrating at the same time ubiquitous computing, user interface design and ubiquitous communication, where work as well as leisure activities were included. In the third geographic area of research in this new field, Japan, researchers, leaded by NRI papers produced by the Nomura Research Institute (leaded that Teruyasu Murakami) adopted the term "*ubiquitous* networks society", talking about the transition from the old *e-Japan* strategy to the new *u-Japan* strategy. Meanwhile, IBM coined the idea of "pervasive computing". One of the first implementations of this new idea were the RFIC tags (radio frequency identification tags), which enabled chain and inventory management in the private and public sectors.

Very soon appeared some challenges associated to the deployment of AmI: security, privatization of governance, privacy, digital ethics, etc. And since 2000 some synergies appeared between European Information and Communication Program (ICT, as having a catalytic impact in three key areas: productivity and innovation, by facilitating creativity and management; modernization of public services, such as health, education and transport; advances in science and technology, by supporting cooperation and access to information) and North-America institutions (NAS and DARPA). Consequently, we can see that AmI was born within private sectors and evolved through public spheres until were included also the military. In the last case, it was thanks to the electronic war and the emergence of cybercrimes, which promoted the Cybercrime Convention on November 23rd, 2001 in Budapest [4]. Despite of signing it, USA were involved into the long-term Echelon spying activities over European and worldwide countries that led to an official EU answer with the official Inform of 2001 [5], and very recently public organizations like Wikileaks (since 2006) or ex-analysts like Snowden (in 2013) showed again how an official country, the USA, was committing cybercrimes against its allies as well as its enemies. Trying to solve general population distrust towards AmI, the Commission created the SWAMI project (Safeguards in a World of Ambient Intelligence), seen as a necessary step towards the resolution of social, legal, organizational and ethical issues related to ambient intelligence. They worked identifying the problems of AmI, creating and analyzing the 'dark scenarios' that AmI could cause to and, finally, creating the best roadmaps for the future of AmI.

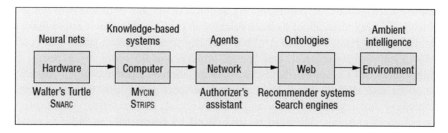

Fig. 1 The evolution of AmI and its relationship with AI. *Source* [8]

1.2 AmI and Artificial Intelligence

There is a strong relationship between computer sciences, Artificial Intelligence (henceforth, AI) and AmI. First, machines were designed to perform basic calculations or basic repetitive mechanical actions; later, was devised a way to create universal programmable machines and very fast their size decreased and became cheaper until make possible its introduction in private homes. The next step was the interconnection of these machines through the web and, as a final era, these machines became so small that were embedded into daily-use artifacts like phones, watches, clothing, and so on: this was the birth of Ambient Intelligence. Here can be found a graphic vision of this process (Fig. 1):

But AmI doesn't work alone or by itself; it needs ways to collect, process and use the ambient data, and this is made by AI mechanisms that make possible an automation of the whole process. Knowledge representation, machine learning, computational intelligence, planning, incompleteness and uncertainty, speech recognition and natural languages, computer vision, robotics or multi-agent systems are fields in which AmI and AI must work together [6]. Soon appeared some ethical aspects related to AmI, some of them related to visible/invisible qualities of AmI systems [7, 8]. Modeling, simulating, and representing entities in the environment were different fields in which Ai and AmI collaborated in order to create better understanding of real human situations. Several technical tools from AI research were necessary to create AmI and at the same time, AI evolved thanks to the specific necessities of AmI.

2 Ambient Stupidity and Some Related User-Designer Pitfalls

2.1 Ambient Stupidity: What Does Is It?

The title of this chapter is 'Ambient Stupidity' (henceforth AS) and it's time to define this concept. By AS I mean *the collection of inefficient devices and support technical instruments or ideas that contributes to an inefficient, bugged or biased ambient intelligence data paradigm.* My definition embraces both bad AmI designs as well as wrong ideas about its main description or goals. My aim is not to make a clumsy and aggressive critic to AmI but to clarify some important points that could help to improve its research and correct implementation. From my expertise in Philosophy of Computer Sciences, Computational Epistemology, Artificial Emotions and HRI [9–20], I have a broad perspective on how interacts nature, technology and cognition (natural as well as artificial) and I consider that there are some important points that must be understood by researchers in order to create better AmI.

At the same time, the idea of 'Ambient Stupidity' must be understood within the framework of recent cognitive theories, such as bounded rationality, or situated cognition. Despite of historical attempts to create a pure superhuman logical thinking process, like the notion of *mathesis universalis* by Leibniz or Descartes during XVII century and that was followed by logicians like Boole or Frege and finally crystallized into symbolic AI by people like Herb Simon, human beings do not think symbolically. At the same time, our thoughts are related to sensorimotor processes as well as to emotional arousals, drives and goal orientation procedures. The norms that guide human rational inferences are far from those expected or dreamed by philosophers all throughout the history. With the first critics to the classic view of rationality [21–23], together to a main reaction against classic AI [24, 25] and the emergence of social relativistic views inside social studies of science (headed by [26]), or a more historical philosophy of science [27], seemed to explain a feeling of experiencing a 'lost paradise of rationality'. But the truth was that these new approaches about cognition and epistemology reinforced and made stronger the naturalization of knowledge as well as improved dramatically the knowledge about the social processes involved into human cultural evolution. Under the light of this new cognitive paradigm, which sees cognition as a process mediated by own bodies, external devices that extend cognitive processes, and that is modulated by emotional drives, the notion of "Ambient Stupidity" does not seem to be a playing with words activity, but instead of it reflects the existence of the failures that are linked to cognitive processes. And AmI is part of a cognitive processes: sometimes will be a connection between internal thoughts and external intelligent systems that will capture them through voice or body commands, while sometimes these cognitive processes will refer to the own process of the artificial architecture that manage the AmI system. We will also see that with the implementation of techniques that make possible easy captures of brain electrical activities, as well with the implementation of diverse robotic prosthesis, the realm of AmI will reach the own user body.

2.2   Some Common Mistakes

In this section I will provide some examples of wrong conceptual elements present in AmI scenarios. Here are some of them:

1. Information: the notion of 'information' is a very rich and complex one [28], and deals with a precise and somehow stable ordination of bits of data (that can have some teleological purpose, like DNA or programs, or not) that we select or create. At the same time, information has become the key concept to understand contemporary societies, according to United Nations [29]. There is not possible to separate raw data from our ontological and epistemological interpretation from data. The process itself of selecting data refers directly to our models about the data structure and nature, and it affects how we see and understand the world. The obvious corollary of the previous ideas is that there is no a true informational reality: we always are selecting groups of data as representative of certain events, artifacts or processes. The self process of arguing about data is a data driven process guided by provisional epistemic rules, that are, at the same time, informational representations about the human cognitive and cultural nature. This should clear our approach to AmI and the kind of data we are obtaining, because all the debates on AmI deal with how to collect, keep and process data. There are more elements to be taken into account: multi-sources data, big data, statistical approach to data, social data, mechanistic models on data, etc. In the nest sections I'll analyze in more depth these ideas.

2. Ambient: in the next decades, smaller and cheaper devices will be able to capture information from several environments. This holistic and worldwide ubiquity will become the new wave or era in AI and AmI. The Ambient cannot be reduced to a static and precisely localized places like the own home, work places or military installations, and instead of it must be understood as a globally distributed as well as multilocalized process: information is gathered from different devices that are around persons (body sensors—external or internal, geo-localization, about emotional estates,...) and at the same time, people is moving or interacting worldwide through these highly distributed systems, most of times online. Besides, with the increasing implementation of intelligent or autonomous devices, like autonomous cars, the range of entities interconnected will even increase dramatically. All this information will be processed thanks to huge and dynamic databases that will need to create new 'mechanisms models' about reality, far from causal or even basic statistic ones. Last DARPA project towards causal big mechanisms will see its size to increase when more datasets will be included into the analysis process [30], this time, with dynamic data that are evolving through the time. *Big AmI* is in front of us, as a greater challenge that has been AmI in last decades, working within the Big Data paradigm but integrating multilocal and multilevel kind of data.

3. Privacy: very sardonically, when the European Commission informed about Echelon USA spying activities [5], included into the main inform the classic text "Sed quis custodiet ipsos custodes?" by Juvenal (h. 60–h. 130 d. C.), *Sat.* 6, 347. This is a key point when we talk about data privacy, a key point in AmI and Society of Information debates. If national security agencies break the right to privacy whenever they want to all these debates are nonsensical. At the same time there is another important aspect related to the inconsistencies of human mental processes (studied by bounded rationality), that I have coined as the *FreeDoom effect*.

The idea is very simple: users accept uncritically free electronic services but at the same time they will deny any kind of permission about their data in a different scenario. But when they accept these free apps or services, they are doomed from an informational point of view. Consider for example, the long list of permissions required to install Facebook into an Android smartphone: (a) personal information, (b) add or modify calendar events and even send e-mails to calendar event guests without explicit direct permission (c) read text messages (MMS or SMS), (d) change connectivity (connect or disconnect from WIFI), (e) system control: draw over other apps, prevent phone from sleeping, re-order running apps, retrieve running apps, toggle sync on and off, adjust or configure the wallpaper, (f) hardware controls: change audio settings, take audio, pictures and videos automatically and without specific permission, (g) modify or delete data stored in memory areas (SD card,…)…it is a long and shocking list of things that hundreds of millions of Earth citizens allow to do to Facebook through their mobile phones. Similar things happen with Google applications (Gmail,…). This is absolutely crazy! Most of times people are arguing about privacy rights but without considering the real extent of some of their actions and about the ubiquitous nature of informational companies: these rights change from country to country at the same time are out of control of any universal jurisdiction. And even in the case of the Budapest Cybercrime Convention the signatory members of the list do not respect it.

4. Regulation: very recently the Court of Justice of the EU (CJEU) found that Spanish citizen Mario Costeja-González had a right to ask Google to remove the links to two 16-year-old newspaper articles about the foreclosure of his home due to unpaid debts (which he subsequently paid) [31]. This has been called the "right to be forgotten". Again, the debate runs uncritically because Google is a private company, not the official record of the reality. Moreover, deep differences among private and public sectors about information rights are expressed not only at local but also global level. With the Internet this process is even more dramatic.

5. Individual versus group. Until now, most of approaches to AmI have been focused on processing single unit sets of data, that is, information from a single unit under analysis (one human being, one house, one building). Despite of the scale of the unit, this approach is insufficient in the light of the Internet of Things [32]. The term "Internet of Things" was popularized by the work of the Auto-ID Center at the Massachusetts Institute of Technology (MIT), which in 1999 started to design and propagate a cross-company RFID infrastructure and now is part of a chain of labs devoted to this process [33]. Despite of the great advancements in this field, there is still a not covered process: how individuals are different when they are considered part of a group. That is, individual versus collective behavior. To be interconnected is not the same as react according to the size of the group and the available communicative tools. At the same time, there are not included future aspects like the introduction of artificial devices, like microchips into human bodies. The foundational researches of Kevin Warwick at Reading University inserting into his body a microchip that allowed

him to move a remote robotic arm or to communicate with his also microchiped wife, are a new level of analysis about how data will be collected and transmitted.

6. Emotional presence: most the capture and cross-analysis of data is still based on abstract measurements of human bodily activity. For strange but also understandable technical reasons, emotional aspects of human performances have not been included into the range of desirable data to be collected. But emotions are the language of mind and regulate in a very determinant way the social life. Social networks have tried to capture part of this language through emoticons, 'like' buttons or similar strategies...but it is still far from the natural expression of emotions. Emotions are a must-have for any future valuable research on AmI. This could help to design better human-friendly machines as well as to create coordinated actions of all the systems that interact via AmI with a human or a group of them.

7. Passive versus (re)active AmI. Most of studies on Ambient Intelligence are based on the gathering of data from human users, with sets of responses but a lack of direct attempt to modify the user attitudes or actions. Other AmI projects can be labeled under 'affordances' category: they enable friendly uses of increasingly technological environments. But I'm talking here about a different question: how to invert the tendency being thus able to modify human actions through specific AmI designs. A good example of this is the recent "Happiness Counter" created by H. Tsujita and J. Rekimoto at U. Tokyo and SONY CSL [34]: one sensor embedded into the fridge, door or a mirror forced users to smile in order to be allowed to use that object. Initially forced to, after some days users tended to produce easily and more natural smiles, and at the same time their psychological welfare improved. This is an example about new ways of creating pro-active AmI environments. Could be of the greatest interest for work places design, conduct control, as well for e-health AmI technologies.

8. Logic and semantic of actions: human actions must be understood as guided by simple natural forces as well as for strategies adopted consciously or not by cultural learning. They are a mixture of natural prewired bodily tendencies as well as the result of special cultural trainings. In both cases, rational processes that execute the decision rules are biased by cognitive malfunctioning (the bounded rationality), by heuristic limitations or by cultural constraints. Therefore, there is not a single logic nor a single semantic for the understanding of human actions and this fact force us to talk about *bounded AmI*, as the real AmI in complex human environments. AmI should consider the cultural differences among users of any technology (even in the same cultural geographical space), as well as temporal constraints (something can be good if it happens or not in the 'correct' sequence order) or personal necessities. AmI will not world in a perfect zero-sum scenario, but into a dynamic, changing, evolving and complex human scenario.

3   What Can Philosophy Do for AmI?

If we look at Classic Eastern as well as Western Philosophies, most of approaches to technology are dystopian and negative. Besides, the interest of most of contemporary philosophers on computer sciences, AI or AmI as crucial aspects of modern societies and the backbone of their thoughts is inexistent. Besides, philosophers, together with psychologists, economists, sociologists, neurologists and some other '-ists', have been researching on rationality. First of all, trying to improve rational thinking, but after some advances, forgot the original conditions in which human thought emerged and dreamed about a 'perfect thinking', beyond mistakes, fallacies or doubt. This led to the classic AI paradigm or GOFAI. But when classic logics showed their own limits and, at the same time, how humans achieved several successful reasoning strategies following strange heuristics, then the notion of 'bounded rationality' emerged as a deep evidence. Humans are special because have big brains (adjusted to the EQ), compute with them a symbolic culture but also because they use continuously emotional background to deal with all these processes. From here, can be understood intentionality, goal-directed actions, ad hoc heuristics that make possible the emergence and evolution of complex actions and theoretical constructs. A possible mistake for AmI developers is to standardize the 'natural' or obvious human reactions in specific monitorized scenarios. The same AmI should be able to evolve, following different strategies to understand the user's actions and interactions, with open methods. Evolutionary computing techniques could be applied into these domains. I'm thinking for example, in real-time haptic AmI employed in real combat scenarios where automatically could decide the best selection of information in order to provide the best choice options for their users. As an specific example I'm imaging on intelligent radar sensors updated with real-time satellite image analysis which could provide pressure information on soldier's torsos thanks to vests equipped with pressure motors distributed among its surface. Therefore, soldiers could feel into their bodies the information about proximity and number of enemy forces, as well as more visual information about armory and related data (on intelligent glasses with augmented reality). The system could be individually operated or shared socially, allowing to an external intelligent system to assist them into the combat best strategies.

At the same time, philosophical approaches can offer to AmI's experts important data about how human act and think, as well as to point to some ethical new debates, following the classic 4 areas of *identity, privacy, security* and *trust*. Some examples of possible contributions or points of view:

- Right to be *AmI blind*: it will be an obvious fact that a great number of users will ask for the right of being not continuously surveilled by the sensors they use as well as for any Big Brother [35]. Surely, this protection will not be a functionality of the host AmI system, but will be a counter-system that any user will have to be equipped with in order to maintain her/his ambient blindness. Here will emerge a vital point: will have a user the right to keep his ambient blind? Do governmental authorities will have the right to force citizens to be open to these technologies? Still today most of experts think on AmI as locally-based systems: offices, houses, buildings…but the Internet of Things aplus ubiquitous computing will make possible the universalization of AmI areas: in fact, the world itself will become an augmented world thanks to a globally interconnected AmI. The cyphering of human activities will be a strongly debatable topic.

- Manipulation or cheating to AmI: there is a second option not considered by all studies, that it, people cheating actively to the AmI sensors. Nobody enjoys being surveilled continuously, but at the same time, there is an extended double-moral among human beings of worldwide cultures. In some cases, this control is avoided arguing about the necessity of security: politicians trying not to be tracked, actresses and actors claiming for their privacy, and so on.
- The previous point is strongly related to the debate: "digital versus real self/identity". I it is a matter of fact that people act differently when they are observed than when they consider themselves not being observed. Humans use masks to hide their inner feelings and motivations. As a counterexample Asperger syndrome patients show extremely sincere attitudes, something that difficult their social life, because humans need to hide and to lie about their real feelings and ideas in order to preserve a social sphere. We want to be studied as a necessary condition for the correct answer to our necessities but at the same time, a completely self-transparency can distort our social life. A last basic scenario: a husband is surfing looking for pornography[1] when his wife is not at home, but when she returns the husband do not want that automatic services show these images at the TV or shared computer facilities.

  – Here must be also considered the notion of individual versus collective, nothing absolutely universal, but a result of our cultural decisions [21, 36]. Good adaptation to these cultural differences will improve AmI systems.

- FreeDoom effect: explained previously, but I insist now again. Most of Internet (and smartphone) users are against their daily life control by policy, street cameras or personal data from public institutions but at the same time they are intensive users of free services like Google (Gmail, Handouts, Google+) or Facebook. These companies collect, store, process and sell the data obtained by the analysis of the behavior of their users. This information is even accessed by governmental agencies, under official request of national security.
- AmI and proxemics: several authors have studies from classic anthropology to modern HRI relationships [37] the meaning of proxemics. Proxemics studies a very important aspect of non-verbal communication: the man's use of space as a specialized elaboration of culture. It explains the personally and culturally

---

[1]Today, there are 420 million pages of internet pornography, and 68 million daily pornographic search engine requests constitute 25 % of total search engine queries, 35 % of all internet downloads are pornographic and 20 % of men admit to watch porn at work. It is a hidden industry, but very important. See: http://dailyinfographic.com/the-stats-on-internet-pornography-infographic, accessed on June, 22nd 2014.

biased way of interpreting distances: by these mechanisms we consider some-
body too much close to us according to her/his relationship (a husband can be at
15 cm from your face for several minutes, but you never would agree to the
same with an unknown person). The point is that AmI respects and breaks and
the same time the classic notions of human proxemics language: because of its
invisibility, most of data capture AmI devices are not perceived by the user; but
this does not imply that the system is crossing all the several natural (and
cultural) barriers with which we are used to. This can be accepted if the AmY
system follows the will of the user, but can be felt as a violation of the intimacy
when the system reacts improperly to the demands of the user, who then realizes
about the loss of privacy regarding all her/his body and space.

- Unknown/unexpected data integration as a loss of control: companies share and
  integrate their data repositories without specific consent or knowledge of their
  users. For example Facebook bought Whatsapp or Google bought Dropcam and
  Nest, among other services. The user is not able to understand and control the
  scalability of their profile.
- Freedom to be wrong or to share specific cultural behaviors: intelligent fridges
  will warn us about expired food, but perhaps we do not want to hear these
  warnings every day despite of the good and healthy intentions of the system. At
  the same time, a 'bad' food can be a necessary ingredient for a specific cultural
  meal.
- Risk assessment adoption in AmI: only naïve approaches to risk assessment can
  consider its use as a definitive solution to AmI problems, because risk assessment is a big controversy field with deep divergences among specialists, institutions or scientific traditions from different countries. Epidemiological debates
  are a good example of this point. If we want to integrate risk assessment into
  AmI studies it must be done with a good epistemological approach [38].
- Embodied AmI: cyborg integration is closer as soon as new tools are available,
  integrating physical bodies with ambient spaces. Kevin Warwick and his projects Cyborg 1.0. and 2.0. (Reading University) were first steps towards cyborg
  technologies. Neil Harbisson makes new researches to solve his achromatopsia
  (although the data are still only available for his brain, not for surrounding
  devices), but at the same time he has included more options once the deice war
  surgically implemented into his skull (like phone call receiving and wireless
  communication). In the case of Harbisson, he has admitted that his device can be
  hackable and even in that possible scenario, his prosthesis has not security
  password access. Zoe Quin, a videogames programmer has recently been
  implanted a NTAG216 chip with NFC technology that allows her to
  block/unblock her smartphone as well as to control access and some uses of a
  videogame (Deus Ex.). This is a real case of Embodied AmI, a research field that
  requires from deep analysis: medical, ethical, cognitive, economic, political, etc.
- Hackable bodies: humans are increasingly living in more intelligent and invis-
  ible environments. The present technologies led us inexorably to the intangi-
  bility of Ambient Intelligence (AmI) frameworks in which brains cannot imitate
  other brains performing bodily tasks. Only users use their bodies inside AmI or

augmented reality environments. There, humans feel themselves understood only because these technologies have been designed to be one step advanced to basic necessities of the users. Anyhow, humans can be confused by one crucial aspect: their interaction with these intelligent systems is not placed under a mutual and symbiotic sensorimotor or emotional context. When these technologies fail, and it can happen very easily, the users have no ways to understand what is happening nor have a natural, from a cognitive perspective, way to connect again with those fake intelligences. If this happens with embodied technologies like medical prosthesis integrated into human bodies (like artificial pacemakers), that can also be hackable, then terror forms part of the existence. But it is not a dark fear that is previsible, emerges silently but destroys everything. Hidden tsunamis of data, of loss of energy, of systems malfunctioning become the expression of a clean and white fear. The most dangerous, because it emerges violating all our natural ways to understand that something is going wrong. Though this fear is embedded into intelligent systems failures, on the other side wonder is also present. Hackable bodies open the possibility of creating a better you. A "you" that can intervene the perception and interpretation of the world, through synthetic emotions or by hacking our multisensory inputs. With the skills of a direct intervention into the reign of physical bodies or personal intimate space (from an anthropological proxemics perspective) of the spectator, the electronic hacktivist will be able to create surprise and a great range of emotional reactions. Cognitive neurosciences have demonstrated during last decades that the crucial aspects of sensorimotor processes are related to cognitive processes.

4   End Remarks

AmI is clearly the next step into the evolution of the integration in daily life of computer technologies and AI products. Despite of the revolutionary nature of AmI, its analysis and media attention was big in the middle of 21st century decade, but decreasing since then. And this is a mistake, because the new social tendencies as well as the advancements in technological equipment and data procession techniques are allowing a next step into the advancement of AmI. Some general mistakes have been pointed and a broad philosophical analysis framework has been suggested in order to prepare and improve the advent of *Big AmI*. At the same time, some critical remarks on the nature of human cognitive processes (the *FreeDoom effect*, or the bounded rationality, among others) have been introduced or reviewed. Finally, the request of new tools to deal with multivariate and dynamic sources of data has been showed as a necessity of future researches. The scenarios that will lead to better roadmaps must be founded on the existence of humans performing bounded rationality processes. As a plausible evolutionary trait of contemporary societies, we need also to take into account the embodiment of AmI technologies as a future step into human social evolution. The powerful philosophical tools that we found into the Western and Eastern traditions can provide us new ways to analyze and improve AmI technologies and models. The sum of the previously explained ideas and critics will allow to surpass 'Ambient Stupidity' and to reach a new and powerful AmI, more adapted to the reality users, considered them as individuals at the same time that socially networked.

Acknowledgements This work was supported by the TECNOCOG research group (at UAB) under the project "Innovación en la práctica científica: enfoques cognitivos y sus consecuencias filosóficas" (FF2011-23238). I thanks specially to Dr. Kiran Kumar Ravulakollu for his interest on my research and confidence on my epistemic contributions. I must also thank to Florence Gouvrit, an electronic artist and professor at Ohio University for her recent discussions about hackable bodies, which have been here sketched.

References

1. Augusto, J.C., McCullagh, P.: Ambient intelligence: concepts and applications. Int. J. Comput. Sci. Inf. Syst. **4**(1), 1–28 (2007)
2. Friedewald, M., Vildjiounaite, E., Wright, D.: Safeguards in a world of ambient intelligence (SWAMI). In: European Report on Ambient Intelligence and the Conference. http://is.jrc.ec. europa.eu/pages/TFS/documents/SWAMI_D1_Final_001.pdf (2006). Accessed 2 June 2014
3. Kuhn, T.S.: The Structure of Scientific Revolutions. UCP, Chicago (1962)
4. Convention on Cybercrime. http://conventions.coe.int/Treaty/EN/Treaties/Html/185.htm (2001). Accessed 2 June 2014
5. Report on the existence of a global system for the interception of private and commercial communications (ECHELON interception system) (2001/2098(INI)). http://www.europarl. europa.eu/sides/getDoc.do?type=REPORT&reference=A5-2001-0264&format=XML&language= EN (2001). Accessed 2 June 2014
6. Ramos, C.: Ambient intelligence—a state of the art from artificial intelligence perspective. In: EPIA'07 Proceedings of the Artificial Intelligence 13th Portuguese Conference on Progress in Artificial Intelligence, pp. 285–295. Springer, Heidelberg (2007)
7. Philips Research Home—AmI: http://www.research.philips.com/technologies/ambintel.html (2014). Accessed 3 June 2014
8. Ramos, C., Augusto, J.C., Shapiro, D.: Ambient intelligence—the next step for artificial intelligence. IEEE Intell. Syst. **23**(2), 15–18 (2008)
9. Crutzen, C.K.M.: Invisibility and the meaning of ambient intelligence. Int. Rev. Inf. Ethics **6**, 53–62 (2006)
10. Vallverdú, J.: Computational epistemology and e-science. A new way of thinking. Mind. Mach. **19**(4), 557–567 (2009)
11. Vallverdú, J., Casacuberta, S.D.: Handbook of Research on Synthetic Emotions and Sociable Robotics: New Applications in Affective Computing and Artificial Intelligence. IGI Global Group, Hershey (2009)
12. Vallverdú, J., Shah, H., Casacuberta, D.: Chatterbox challenge as a testbed for synthetic emotions. Int. J. Synth. Emotions **1**(2), 57–86 (2010)
13. Vallverdú, J., Casacuberta, D.: The game of emotions (GOE): an evolutionary approach to AI decisions. In: Ess, C., Hagengruber, R. (eds.) The Computational Turn: Past, Presents, Futures? pp. 158–162. Proceedings IACAP 2011. MV-Verlag, Münster (2011)
14. Vallverdú, J. (ed.): Creating Synthetic Emotions through Technological and Robotic Advancements. IGI Global Group, Hershey (2012)

15. Vallverdú, J.: The meaning of meaning: new approaches to emotions and machines. Aditi J. Comput. Sci. 1(1), 25–38 (2013)
16. Vallverdú, J.: Ekman's paradox and a naturalistic strategy to escape from it. IJSE 4(2), 7–13 (2013)
17. Vallverdú, J.: 6A. Jordi Vallverdú on Muehlhauser and Helm's (The singularity and machine ethics). In: Eden, A.H., Moor, J.H., Søraker, J.H., Steinhart, E. (eds.) Singularity Hypotheses. A Scientific and Philosophical Assessment, pp. 127–128. Springer, Germany (2013)
18. Vallverdú, J., Casacuberta, D., Nishida, T., Ohmoto, O., Moran, S., Lázare, S.: From computational emotional models to HRI. Int. J. Rob. Appl. Technol. 1(2), 11–25 (2013)
19. Vallverdú, J.: Artificial shame models for machines? In: Lockhart, K.G. (ed.) Psychology of Shame: New Research, pp. 1–14. Nova Publishers, NY (2014)
20. Casacuberta, D., Ayala, S., Vallverdu, J.: Embodying cognition: a morphological perspective. In: Vallverdú, J. (ed.) Thinking Machines and The Philosophy of Computer Science: Concepts and Principles. IGI Global, Hershey (2010)
21. Ashton, K.: That 'Internet of Things' Thing, in the real world things matter more than ideas. RFID J. (2009). Accessed 3 June 2014
22. Advancing the Internet of Things for Global Commerce: http://www.autoidlabs.org/ (2014). Accessed 3 June 2014
23. Happiness Counter: http://vimeo.com/29169237 (2014). Accessed 3 June 2014
24. Nisbert, R.E.: The Geography of Thought. The Free Press, NY (2003)
25. Walters, M.L., Dautenhahn, K., Te Boekhorst, R., Koay, K.L., Syrdal, D.S., Nehaniv, C.L.: An empirical framework for human-robot proxemics. In: Proceedings of New Frontiers in Human-Robot Interaction: Symposium at the AISB09 Convention, pp. 144–149
26. Vallverdú, J., Delgado, M.: Values in controversies: stem cell research. Bio-Phronesis: Revista de Bioética y Socioantropología en Medicina 4(2), 1–27 (2009)
27. Wright, D., et al.: Safeguards in a World of Ambient Intelligence. Springer, Germany (2008)
28. Casacuberta, D., Vallverdú, J.: E-science and the data deluge. Philos. Psychol. 27(1), 126–140 (2014)
29. Tversky, A., Kahneman, D.: Judgment under uncertainty: heuristics and biases. Science 185, 1124–1131 (1974)
30. Stich, S.: Could man be an irrational animal? Some notes on the epistemology of rationality. Synthese 64, 115–135 (1985)
31. Gigerenzer, G.: Bounded and rational. In: Beckermann, A., Walter, S. (eds.) Philosophie: Grundlagen und Anwendungen, pp. 203–228. Mentis, Paderborn (2008)
32. Brooks, R.A.: Elephants don't play chess. Rob. Auton. Syst. 6, 3–15 (1990)
33. Brooks, R.A.: Intelligence without representation. Artif. Intell. 47, 139–159 (1991)
34. Adriaans, P.: Information. In: Zalta, E.N. (ed.) The Stanford Encyclopedia of Philosophy. http://plato.stanford.edu/archives/fall2013/entries/information/ (2013) (Fall Edition). Accessed 2 June 2014
35. UN: World Summit on the Information Society (WSIS). http://www.unesco.org/new/en/communication-and-information/resources/multimedia/photo-galleries/world-summit-on-the-information-society-wsis/ (2003–2014). Accessed 2 June 2014
36. Google and the "right to be forgotten": http://www.wired.co.uk/news/archive/2014-05/15/google-vs-spain (2014). Accessed 3 June 2014
37. Latour, B.: Visualization and cognition: thinking with eyes and hands. Knowl. Soc. 6, 1–40 (1986)
38. DARPA Big Mechanism: http://www.darpa.mil/Our_Work/I2O/Programs/Big_Mechanism.aspx (2014). Accessed 2 June 2014

Author Biography

Jordi Vallverdú Ph.D., M.Sci., B.Mus, B.Phil is Tenure Professor at Universitat Autònoma de Barcelona (Catalonia, Spain), where he teaches Philosophy and History of Science and Computing. His research is dedicated to the epistemological and cognitive educational aspects of Philosophy of Computing and Science and AI. He is Editor-in-chief of the International Journal of Synthetic Emotions (IJSE), and as researcher is member of the IACAP, Convergent Science Network of Biomimetic and Biohybrid systems Net member, member of the Spanish Society of Logic, Methodology and Philosophy of Science, member of the GEHUCT (Grup d'Estudis Interdisciplinaris sobre Ciència i Tecnologia) research project, member of the TECNOCOG (Philosophy, Technology and Cognition Research Group), member of EUCogIII, Main researcher of SETE (Synthetic Emotions in Technological Environments), and Expert of the Biosociety Research (European Commission). He has written 8 books as author or editor: (2009) Bioética computacional, México: FCE.; (2009) Handbook of Research on Synthetic Emotions and Sociable Robotics: New Applications in Affective Computing and Artificial Intelligence, coedited with D. Casacuberta, USA: IGI Global Group; (2009) Proceedings of the VIIth European Conference on Philosophy and Computing, Editor, UAB: Bellaterra; (2010) Thinking Machines and the Philosophy of Computer Science: Concepts and Principles, Editor and author, USA: IGI Global Group; (2011) ¡Hasta la vista Baby! Un ensayo sobre los tecnopensamientos, BCN: Anthropos; (2012) Creating Synthetic Emotions through Technological and Robotic Advancements, Editor and author, USA: IGI Global Group; (2015) Bayesian versus Frequentist Statistics, Springer, in press. In 2011 he won a prestigious Japanese JSPS fellowship to make his research on computational HRI interfaces at Kyoto University. He was keynote at ECAP09 (TUM, München, Germany), EBICC2012 (UNESP, Brazil) and SLACTIONS 2013 (Portugal). See more at: http://orcid.org/0000-0001-9975-7780#sthash.rhvZw6l3.dpuf.

Security Implementations in Smart Sensor Networks

Mohamed Fazil Mohamed Firdhous

Abstract Wireless sensor networks have become one of the widely deployed networking technologies in the recent times due to the capabilities and advantages of them. The applications of wireless sensor networks include many civilian and industrial applications to military applications. Due to the distributed nature of these networks, deployment in remote and open areas and many constraints in individual nodes, these networks are vulnerable to several security threats. Many security mechanisms and algorithms proposed for the implementation in the traditional networks cannot be implemented in wireless sensor networks due to the unique nature of these networks and nodes. Many active research programmes have been carried out throughout the world for making wireless sensor networks more secure and user friendly. This chapter takes an in-depth look at some of the prominent mechanisms, schemes, algorithms and protocols published in the literature.

1 Introduction

Smart sensor networks have found a place in many popular application domains especially for monitoring, tracking and control purposes [1]. A sensor network is an array of sensors and other nodes interconnected by a network for the purpose of transmitting the data captured and other information between these nodes. In these networks, the sensors occupy the main position as they play the important role of capturing the information that is considered to be of value when processed. With the advancement of semiconductor and sensor technologies, smart sensors have been developed that can carry out many more tasks than just capturing the data. Smart sensors are required to have seven major elements in them. They are namely sensor, signal conditioner, analog to digital converter, application algorithms, data storage area, user interface and communication interface [2]. These additional elements make these sensors to be more versatile, reliable and secure while requiring less

M.F.M. Firdhous (✉)
Faculty of Information Technology, University of Moratuwa, Katubedda, Sri Lanka
e-mail: Mohamed.Firdhous@uom.lk

© Springer International Publishing Switzerland 2016
K.K. Ravulakollu et al. (eds.), *Trends in Ambient Intelligent Systems*,
Studies in Computational Intelligence 633, DOI 10.1007/978-3-319-30184-6_8

maintenance compared to normal sensors. Smart sensors can be setup fast and have the capability of reprogrammed to suit the changes in requirements. Also these sensors can be monitored remotely, this eases the administration of these networks to a very great extent [3].

With the increased deployments and applications of sensor networks, many issues that demand immediate and special attention have also come to the fore. One such major issue demand the critical attention of the implementers as well as researchers is security [4]. Security in smart sensor networks not only need to be enhanced but also made to be more rugged in the face of increased security threats and new methods of attacks. The security in sensor networks must be addressed from multiple directions requiring a multi-pronged approach. The areas that require special attention can be summarized as: security of the sensor nodes, security of the information transferred and security of the information path. Implementing security in sensor networks is a challenging task due to inherent constraints in the wireless sensor networks such as remoteness of implementation, limitations in processing power, instability of the network and shortage of energy supplies [4–8].

This chapter presents an in-depth evaluation of security implementations in smart sensor networks, specifically on three main areas. They are namely: security of smart sensor nodes, security of data transferred and security of routing in smart sensor networks. The evaluation primarily concentrates on the present security implementations with special reference to their principles, strengths and weaknesses along with the future directions of research in these specific aspects.

2   Smart Sensor Networks

A sensor network is an array of sensors possibly of different kinds and processors that are interconnected by a communication network for the purpose of transferring data and control information between them [9]. A sensor can be of single modal or multi modal depending the requirement and the complexity of the sensor itself. A single modal sensor can carry out only one sensing function and made of a single technology. On the other hand, multi-modal sensors are multifunctional and may be composed of many sensing hardware created using optical, acoustic, chemical, infrared, magnetic, seismic, tactile, temperature, gravity, pressure, electric, semiconductor etc. In recent times, semiconductor sensors have become more popular due to their functionality and versatility [10]. For example, modern semiconductor gas sensors can detect more than 150 gases making them the most preferred choice in many industries like automotive, consumer, commercial, industrial, indoor and outdoor air quality monitoring and environmental monitoring [11]. Also, semiconductor sensors have special characteristics such as better sensitivity, faster response time, long term stability and longer life time compared to other sensors.

Smart sensor networks can be created installing intelligence into the sensors or closer to them [9]. When the processing capability along with sensing and other required units such as signal conditioner, analog to digital converter, application algorithms, memory for data and application storage, user interface and communication interfaces are built into a single module, it is known as a smart sensor [2]. When intelligence is integrated into an aggregator node that receives raw data from neighbouring not so smart sensor nodes and processes them before sharing it with other aggregator nodes in the network or a central processing unit, the intelligence or smartness is located closer to the nodes. Thus the aggregator nodes are considered to be more capable and powerful compared to the other simple nodes in the network. Simple nodes just broadcast the data they collect while the smart nodes process them for the purpose of extracting information through various operations such as validating, deriving, integrating etc., before transmitting. Since the data is validated and processed closer the source itself, it saves the valuable network bandwidth and in sometimes energy by not transmitting invalid or partial information.

Smart sensor networks can be deployed for various purposes such as monitoring the environment, functions and operations of machinery or the human body itself or movement of objects within certain premises or operations [12]. Depending on the type of application and the type of nodes deployed, these networks will have various capabilities and limitations. Depending on the type of connection between the nodes, sensor networks can be divided to two categories known as wireless sensor networks and wired sensor networks. Wireless sensor networks suffer from many limitations compared to wired sensor networks due to their inherent nature. The main limitations of wireless sensor networks include limited power, limited processing capabilities within nodes and unstable communication between nodes. Generally wireless sensor networks are also implemented far away from the final processing centres in remote locations making the management of these sensor nodes a difficult task.

2.1 Sensor Node Placement

Sensor node placement is an important aspect that must be given proper consideration for the successful implementation of sensor network [13]. Sensor nodes can be either placed deterministically or randomly depending on the type of application, size of deployment, number of nodes to be placed and the geographical area to be covered. In industrial applications, sensors are placed deterministically at strategic points for collecting the right information. Generally in industrial settings, it is the operation and functions of machineries and related equipment are monitored. When the health of a machine is monitored in an industrial setting, the sensors are placed at various points within the machine or closer to the machine for monitoring the temperature, flow of coolants, properties of coolants etc. When indoor environments or outdoor environments are monitored in a limited fashion like traffic monitoring system or the monitoring of pollutants in a certain area, the sensors are placed in a deterministic manner.

When sensors are placed for monitoring a large area for environmental changes, aftermath of natural disasters or military operations, it is not possible to place them deterministically due to the large number of nodes to be placed or the accessibility issues in these areas [14]. Generally during large scale sensor deployment in a geographically distributed manner, sensors are placed randomly by dropping them off from an airplane or some other method [15]. This kind of placements have many shortcomings including coverage and communication problems. When sensors are dropped randomly, certain areas may have been deployed with many nodes resulting in coverage overlaps and wasting of resources. On the other hand the areas, where there are insufficient nodes, would have coverage holes and connectivity problems resulting in inefficient monitoring and isolation of sensor nodes. Hence nodes must be placed in an efficient and effective manner to reduce the problems arising from coverage overlaps, holes and communication. In many situations, redundant nodes are deployed in order to overcome the problems of shortage of coverage and communications in random node deployments [16].

2.2 Sensing and Data Acquisition

The set of nodes deployed in a particular application can be either homogeneous or heterogeneous. When all the nodes deployed are of the same type and have similar capabilities, it is known as a homogeneous deployment. In a heterogeneous deployment, certain nodes may have different capabilities compared to other nodes used. One of the main attribute that is used for categorising nodes in a deployment is their sensing range. The sensing range is the area across which a node is capable of detecting the presence or absence of an object or phenomenon. Certain types of nodes may have different sensing ranges and can choose a specific range out of all the available ranges as its working range depending in the requirements. A general assumption is that when a large sensing range is used by a sensor node, it consumes more energy. In heterogeneous deployment, the nodes with larger sensing ranges are generally used as cluster heads due to their advanced capabilities [17]. In remote deployments of wireless sensor networks, the total amount of energy available in nodes will determine the life of the networks. Hence it is recommended to use the minimum amount of energy for all sensor operations including sensing, processing and communication in order to prolong the life of sensor networks. Ranjan and Kar [18] have provided a method for determining the optimal number of cluster heads for homogeneous sensor networks using reasonable energy consumption model.

2.3 Connectivity in Wireless Sensor Networks

The other important parameter that affects the performance of a sensor network is communication range. In a multi-hop sensor network, communication nodes are linked by a wireless medium such as radio, infrared, or optical media [19]. Once the data has been collected, that data needs to be transmitted to the processing centre.

Connectivity between nodes is important to ensure that every sensor node can communicate with the processing centre [20]. In a multi-hop wireless sensor network, the network is said to be fully connected if every pair of nodes is able to communicate with each other, either directly or via intermediate relay nodes.

A sensors network is considered to be connected, only if there is at least one path between each pair of nodes through which successful communication can take place. Hence for the successful transfer of data from any given node to the processing centre, there must be a communication path from that node to the processing centre. Connectivity between nodes depends primarily on the existence of paths and affected by changes in topology due to mobility, failure of nodes and attacks that cause loss of links, isolation of nodes or partitioning of the network [21]. Though the cost of individual sensor is relatively low, the total cost of implementing a sensor network could be high due to the large number of sensor nodes required to setup a network. Therefore, it is important to find the minimum number of nodes required for a wireless sensor network to achieve full connectivity while optimizing coverage at the same time.

Since nodes in wireless sensor network are connected to other nodes via radio, infrared or optical media, the communication range of the nodes will determine whether the nodes still are part of the network or have become isolated from other nodes. When a sensor node needs to communicate with other nodes, it must be within the communication ranges of both transmitting as well as receiving nodes. The communication range of a sensor node is generally determined by the transmit power, receiver sensitivity and the total attenuation introduced by the transmission path. The relationship between the communication range and sensing range of sensor nodes for maintaining communication and proper coverage is given by Formula 1 [22].

$$R_c \leq R_s^1 + R_s^2 \tag{1}$$

where R_c is the communication range of the nodes and R_s^1 and R_s^2 are sensing ranges of node 1 and node 2 respectively.

The relationship given in Formula 1 can be better explained graphically as shown in Fig. 1.

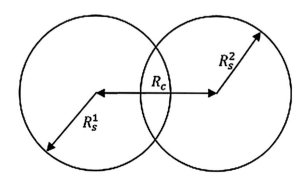

Fig. 1 Relationshipe between communication range and sensing ranges

It must also be noted that maintaining larger communication range require more energy. Hence when deciding an optimum distribution of sensor nodes in a large network, many things including connectivity, coverage, energy consumption, cost and flexibility need to be taken into account.

2.4 Communication Protocols

The successful operation of a wireless sensor network largely depends on the communication protocol chosen for the implementation [7]. The communication protocol chosen every aspect of the wireless sensor network including architecture, data rate, network size, span, power management and security. One factor that is common to all the available communication protocols is that they all are low power communication protocols. Currently there are three communication protocols that can be chosen for the implementation in wireless sensor networks. They are namely Bluetooth, IEEE 802.15.4 and ZigBee [23–25]. The following subsections briefly discuss these protocols with special reference to their suitability for the implementations in wireless sensor networks when security is the main concern.

2.4.1 Bluetooth

Bluetooth that has been standardized by IEEE 802.15.1 has been initially developed and standardized for the low power wireless devices [23]. The design of Bluetooth requires all nodes within its piconet to be synchronized within a few microseconds. This requirement cannot be met by many wireless sensor networks as they have large network latencies due to several constraints within them [26]. With typical Bluetooth configuration, it would take around 2.4 microseconds to establish a connection. Also typical Bluetooth radios consume hundreds of milliwatts power just for monitoring the channel. All these shortcomings makes Bluetooth unsuitable for implementation in wireless sensor networks.

2.4.2 IEEE 802.15.4

The IEEE 802.15.4 was developed for low rate wireless personal are networks [24]. Wireless personal area networks require little or no infrastructure at all for successful implementation and operation. IEEE 802.15.4 allows the implementation of small, power efficient and inexpensive solutions using a wide range of devices. The features of IEEE 802.15.4 allows the realization of the objectives of personal area networks, that are ease of installation, reliable data transfer, short range operation, low cost of implementation and maintenance and reasonable battery life. These are the main objectives of wireless sensor networks as well, hence IEEE 802.15.4 is very suitable for the implementation in wireless sensor networks. IEEE 802.15.4 standard defines both physical and media access control layers along with component devices and supported network topologies. There are many security suits defined in this standard.

At the basic level, it is possible to either enable or disable security. Security can be disabled by enabling the unsecured mode, which selects the null security suit. An application can select the appropriate security level by entering the required parameters in the radio stack. If no parameter is entered then, no security is enabled by default.

A link layer protocol provides four basic security services. These are access control, message integrity, message confidentiality and replay protection. Access control is enabled through an access control list. The access control list enables message filtering for accepting messages only from selected nodes in the list. Message integrity and authentication is achieved through a message authentication code appended to every frame of data transmitted. Message authentication code is computed using a secret cryptographic code shared by both sender and receiver. When a data frame is received, the receiver recomputes the message authentication code using the cryptographic key in its memory and checks it against the message authentication code received with the frame. If the message authentication codes match with each other, then the data is accepted as genuine, otherwise it is discarded. Without compromising the secret key, it is impossible for an adversary to change valid messages or introduce phoney message into the network. Sequential freshness checks carried out on each received frame enables the detection of replay attacks. The receiver maintains a received frame counter for every message sequence received. When a frame with a counter value equal to or less than the sequence counter value stored in the memory, it is discarded as a duplicate or replay frame.

The access control and message integrity checks can effectively eliminate the unauthorised parties from sending messages and participating in network activities. The authorised nodes can easily detect the messages from rogue nodes initially by filtering messages by the access control list. Even when a rogue node fakes the identity of a genuine node, the message authentication code will help the authorized node to detect the phoney or compromised message frame and discard it.

There are eight different security suites defined within the IEEE 802.15.4 standards. Based on the type of functionality provided, these suites can be broadly categorised into four different groups as shown in Table 1. The broad categories are namely, no security, encryption only, authentication only and encryption and authentication.

Encryption is performed by Advanced Encryption Standard (AES) algorithm. The united States government has accepted AES algorithm as its official standards for its organizations to protect sensitive information [7]. Counter mode cryptographic operation with AES (AES-CTR) uses AES as the block cipher providing access control and encryption along with optional sequential freshness.

Table 1 Security suites defined in IEEE 802.15.4 [7]

Name	Description
Null	No security
AES-CTR	Encryption only—CTR mode
AES-CBC-MAC-128	128 bit MAC
AES-CBC-MAC-64	64-bit MAC
AES-CBC-MAC-32	32-bit MAC
AES-CCM-128	Encryption and 128 bit MAC
AES-CCM-64	Encryption and 64 bit MAC
AES-CCM-32	Encryption and 32 bit MAC

Message authentication is carried out by cipher block chaining with message authentication code (CBC-MAC). Message authentication code is created using block cipher in CBC mode over the entire data packet including the length of the authenticated data. The detailed description of this process is included with the IEEE 802.15.4 standard itself. Depending on the level of security required, it is possible to select 128, 64 or 32 bit message authentication codes within this mode.

AES-CCM mode provides both authentication and encryption for better security. This mode of operation requires three inputs. They are namely; the data payload to be encrypted and authenticated, the associated data along with the headers to be authenticated only and nonce to be assigned to the payload and associated data.

2.4.3 ZigBee

ZigBee is an industrial consortium setup for the purpose of developing a standard data link communication layer for ultra low power wireless communications [25]. Instead of building from scratch, ZigBee standard has been built on top of IEEE 802.15.4. ZigBee network layer has been designed to work above the physical and media access control layers defined under IEEE 802.15.4 standard. The ZigBee network layer functions include mechanisms for joining and leaving a network, apply security to frames, routing the frames to the intended destination and extra security services including key exchange mechanisms and authentication beyond IEEE 802.15.4.

ZigBee specification introduces a new concept known as "trust centre" played by the ZigBee coordinator. The trust centre controls and administers other devices that are willing to join the network and distributes the appropriate key information among them. The trust centre is entrusted to play three specific roles with respect to managing security in the network. They are namely, trust manager, network manager and configuration manager. The trust manager authenticates the devices that apply to join the network. The network manager maintains and distributes the keys among the members of the network. The configuration manager's task is to enable end-to-end security between devices.

A ZigBee enabled network can work in distinguished modes known as residential mode and commercial mode. In the residential mode, where low security application are run, only device authentication is carried out prior to joining the network. No keys are distributed in the residential mode operation persevering much of the memory for data processing operations. On the other hand, the commercial mode is intended for use in high security environments that require not only the authentication of devices but also managing the integrity of information transferred. In commercial mode, the trust centre first authenticates the devices, distributes the keys among them and maintains freshness counter for every device in the network. This enables centralized control and management of keys. Central management with a single trust centre may not scale well with large networks with hundreds or thousands of devices as the memory requirements for managing large number of keys and updating them regularly will be prohibitively high. This shortcoming can be easily overcome by dividing the network into small clusters and managing the keys locally.

ZigBee security services use three types of keys known as master keys, link keys and network keys. The master key that is installed first in the factory or out of band is responsible for long term security between devices. on the other hand link keys and network keys are basis for security between devices and the entire network respectively. The link and network keys employ symmetrical key-key exchange handshake between devices.

3   Security Challenges in Smart Sensor Networks

The nodes deployed in large wireless sensor networks are characterized by their low cost, small size and resource constraints [8]. These nodes have limited processing capability, storage capacity, communication bandwidth and range, energy and sensing range. Due to these constraints, it is not possible to employ conventional security mechanisms and algorithms in a wireless sensor network. Hence when conventional security algorithms and mechanisms are to be employed in a wireless sensor network, they must be optimized to suit the demands, limitations and the environment in which they are deployed [27]. The main limitations of sensor networks with respect to security are explained in Sect. 3.1.

3.1   Constraints in Wireless Sensor Networks

One of the main constraints in a sensor node is the limited energy available for its operations. Generally sensor nodes are powered by small cells (batteries) of limited capacity that can be exhausted in a short time, if not used wisely [7, 10]. The problem of energy consumption is exaggerated due to the fact these batteries cannot be recharged or replaced once they have been deployed [7]. The energy in a sensor node is consumed in three main parts. They are namely the transducer, transmitter and the microprocessor. It has been found that the amount of energy consumed for transmitting one bit of information is equal to about executing 800–1000 lines of codes in the microprocessor [28]. Hence it can be seen that transmission is much more expensive than processing in a sensor node. When security mechanisms are implemented in traditional manner, they result in the expansion of the messages due to the redundant bit added by the security mechanism. This is very costly for implementation in sensor networks in terms of energy consumption.

Limited storage and memory capacity of sensor nodes is another constraint in wireless sensor networks. The storage area in a sensor node generally consists of flash memory and volatile Random Access Memory (RAM) [29]. The flash memory is used for storing permanent information such as operating system and programme codes, while the RAM can hold the programme codes currently in use, data and intermediate results. Hence the memory of a sensor node hardly has any space in its memory for holding and executing complex security algorithms and applications.

Since the sensor nodes and communication paths are affected by various environmental conditions, the communication in a wireless sensor network may not be as reliable as in a wired network. Due to the less overhead associated with connectionless communication protocols, they are commonly employed in wireless sensor networks [30]. The connectionless communication protocols are inherently less reliable. This reduced reliability in communication provides a haven of opportunity for attacks like sink attack, denial of service attack etc. Packet errors and loss will also play a big role in reduced reliability of these communication paths. Due to the higher error rates and employment of connectionless protocols will further demand error detection and correction mechanisms embedding additional bits further reducing the amount of space available for security implementations.

The other major issue confronting the communication in wireless sensor networks in large latencies from source to destination [26]. Higher latencies in communication path is the result of low bandwidth connections, network congestion, multi-hop communication and processing in intermediate nodes. Higher latencies result in loss of synchronization that is essential in many security implementations such as distribution of cryptographic keys, critical event reports etc. Loss of synchronization may also help attackers engaged in replay attacks where time stamping and timely delivery play an important role in containing these attacks.

Sensor networks use broadcasting as the common mode of transmission instead of directed communication [6]. Broadcasting helps nodes transmit the data to all the neighbours enabling the nodes to find the available end to end path even in the case of unavailability of some nodes on the way. Broadcasting can be easily exploited by adversaries for eavesdropping sensitive information with relative ease. Broadcasting can be used by adversaries to transmit commands and data to nodes by capturing a single node in the network, even if the transmission is secured by a pre-deployed global key.

Many wireless sensor networks are deployed in remote areas where the nodes are left unattended and managed remotely [31]. This increases the likelihood of physical tampering by attackers. Such physical tampering is more difficult detect as well almost impossible to stop due to the remoteness of the implementations. This type of node capture attacks are very serious in nature as compromising the security of a single node can pollute the entire sensor network [6].

Every node in a network must be installed with complete security as any node can be the target of an attack [7]. This demands that the security must pervade every aspect of the design of wireless sensor network design as any component left without security will be easily exploited by an adversary. This is high level of security implementation compared to traditional security implementations in conventional networks [32]. High level of security implementation requires more resources and time to implement making the deployment of wireless sensor network more expensive.

Wireless sensor network protocols heavily depend on application scenarios [33]. Hence generic security mechanisms need to be customized to suit each and every application domain. This puts a heavy burden on application developers and increases the application development cost. If the customizing operation is not carried out taking all the aspects into account or any aspect was overlooked, it may create security threats. Also this kind of mass customization makes it difficult to identify the bugs that can be exploited by the adversary as every implementation is different from each other.

3.2   Types of Attacks

This section briefly describe the possible security threats to wireless sensor networks. With the increase of popularity and development of wireless sensor networks, the number and types of threats and attacks carried out on these networks have also increased [7]. Many of the attacks have been identified and described in [34]. These attacks can be broadly categorized into four main groups. They are namely, attacks against the privacy of network, denial of service attacks, impersonation or replication attacks and physical attacks [7].

Some of the most common attack types are described below:

Selective forwarding: Selecting forwarding involves a malicious node dropping certain messages intentionally, while forwarding only a subset of messages it receives. The malicious node that carries out this kind of attack becomes a preferred intermediate node for unsuspecting source nodes as the forwarded messages undergo low latencies faking a shorter route. The impact of this attack depends on two main factors such as the location of the adversary and the number of packets dropped. When the adversary is closer to the base station, it will attract many more frames that it would normally do, if located far away. More the packets dropped higher the energy saved, as the transmission of packets requires a lot of energy. Hence the malicious node can stay alive longer than a normal node perpetuating its attack.

Sinkhole attack: Also known as black hole attack is where a malicious node attracts the traffic towards a compromised node. Generally this kind of attack is carried out faking a base station by a malicious node. A network with a single base station is more susceptible to this kind of attack.

Sybil attack: In this kind of attack, a malicious node presents multiple illegitimate identities to unsuspecting nodes. The identities presented by a node could be either fabricated ones or stolen from legitimate nodes or both. Once a node assumes many identities, it can launch many different types of attacks such as negative reinforcement, stuffing ballot boxes of a voting scheme such as trust computing etc. Sybil attacks are generally carried out against routing algorithms and topology maintenance.

Wormhole: In wormhole attacks, an adversary placed close to a base station channels the traffic over a low latency link. This effectively creates a sink hole completely disrupting the traffic.

HELLO flood attack: In this attack, the malicious node broadcasts a HELLO message with strong transmission power pretending to be coming from the base station. The nodes receiving this HELLO message would respond to them effectively wasting their energy. The other effect of this kind of attack is that the unsuspecting nodes would forward their messages to this malicious node falsely assuming it to the base station.

DoS attack: Denial of service attacks on a wireless sensor networks can be carried out using various techniques. At physical level, radio jamming by transmitting a more powerful signal on the same frequency or exhausting the battery power are common methods. At other level, the legitimate traffic can be diverted from the intended node or illegitimate traffic diverted towards a genuine node effectively making it unavailable for legitimate traffic.

Traffic analysis attack: It is possible to identify the location of the base station by closely monitoring the network traffic patterns. If an adversary can compromise the security of the base station, the entire network would be affected.

Node replication attack: This attack is carried out by copying the identity of a legitimate network node by a malicious attacker node. The results of this attack would be corrupted, misrouted or deleted packets.

Eavesdropping: Since wireless sensor networks generally employs broadcasting as the mode of communication, it is possible for a malicious node to gather all the information transmitted in the network, if they are not encrypted. Eavesdropping could also be the first step in a more powerful and serious attack such as wormhole or sink hole attack.

Tampering: Since the wireless sensor nodes are generally left unattended in remote locations, it is possible for adversaries to physically tamper them compromising all the security implementations.

Table 2 summarizes and classifies the attacks discussed above into different layers of a communication stack based on where they can possibly be effected.

Table 2 Senor network attack classification

Layer	Type of attack
Physical layer	DoS—jamming, tampering
	Sybil
Data link layer	DoS—collision, exhaustion, unfairness
	Interrogation
	Sybil—data aggregation, ballot stuffing
	Node replication
	HELLO message flood
Network layer	DoS—flooding, spoofing, sink holes
	Sybil
	Wormhole
	Traffic analysis
	Selective forwarding
	Node replication
	HELLO message flood
Transport layer	DoS—flooding, desynchronization
Application layer	DoS—flooding, diversion
	Eavesdropping

4 Security of Smart Sensor Nodes

Generally wireless sensor networks are implemented in remote locations for monitoring various things including the environment, enemy movements in military applications etc. [35]. Compared to conventional network devices, wireless sensor nodes are more susceptible to attack as they are physically accessible by adversaries [7, 8, 34]. Once a sensor node is physically tampered with, the entire security implementation in the node including the cryptographic keys can be compromised. Hence physical security of sensor nodes is of utmost important. Since it is nearly impossible to protect the nodes from physical tampering by adversaries, many schemes have been developed for detecting malicious or tampered nodes and isolating them. This section takes an in depth look at some of the prominent malicious node detection schemes reported in the literature.

4.1 Threats to Wireless Sensor Nodes

Compared traditional network nodes, wireless sensor network nodes face several additional threats due to the very nature of their implementations. Generally, wireless sensor network nodes are located in open space that can be considered as insecure and hostile [8]. When an environment is considered to be insecure or hostile, generally the physical security in the area is beefed up with various special mechanisms such as perimeter security through the implementation of security cameras, personnel and policies. But, due to the open nature of the wireless sensor network implementation, the above mechanism cannot be implemented as it is.

The main threats faced by the wireless sensor nodes deployed in the open area are tampering, theft and physical destruction [8]. These attacks can cause irreversible attacks to the nodes and sometimes to the entire network if not handled properly and curtailed at the beginning itself.

Tampering involves the modification of the sensor from its normal operation. An adversary can get hold of the cryptographic keys installed in the nodes, when he gets physical access to the nodes easily compared to attacking the nodes remotely or through data analysis. Also the attacker can now alter the physical hardware including circuitry and wiring or modify the program code as he wants. In the worst case, the entire sensor node can be replaced with malicious sensor node itself.

Theft and physical destruction of sensor node make them totally unavailable for use by authorised users. Both these attacks fall under the denial of service attacks as they deny the genuine user from using these nodes and getting the intended services from them.

4.2   Security Schemes for Protecting Wireless Sensor Nodes

Since wireless sensor nodes have been installed outdoors open to both genuine and malicious users on the whole, it is difficult or many a time impossible to protect them from the physical damages caused by malicious attackers. Installing the sensor nodes more densely than needed may reduce the impact of theft or physical destruction to the nodes [6, 7]. On the other hand, when a node is tampered with, it must be detected, identified and isolated from the network. Many schemes, mechanisms and protocols have been proposed in the literature for identifying misbehaving nodes. This section takes in detailed look at some of the important mechanisms for identifying and isolating them reported in the literature.

Zia and Zomaya [34] have presented a malicious node detection mechanism based on monitoring its own message retransmitted by a neighbouring node in transit. In this mechanism, the source node first forwards the message to one of its neighbours for the purpose of routing it towards the base station. Once the transmission is completed, the source node converts itself to monitoring node actively observing the retransmitted message by the neighbour. If the retransmitted message resembles the original message, the monitor terminates its task and continues with its normal operations. If the retransmitted message differs from the original message, it updates the locally maintained node suspicious table. Once the number of entries for a node in its suspicious table goes beyond a predefined threshold, it informs its neighbours about the suspicious node. The neighbours then respond back to this message with their own opinion based on their observations. When the suspicious entry for a given node increases beyond a threshold, it is then informed to the cluster head. Cluster head will then isolate the suspicious node as malicious barring all the members from communicating with it and dropping all the messages from the identified malicious node in the future. This mechanism looks robust as node monitors its own message being retransmitted for identifying a malicious node. This mechanism has two main drawbacks. First all the nodes must use the same "link key" for encrypting the message. If different node pairs use different link keys for encrypting the message, it is not possible to identify the modification of the message just by observing the retransmitted message. In such a situation, this mechanism totally fails. Second, this mechanism is prone to collusion attacks, as the opinion of neighbours about a suspected node are taken in without any further inquiry or clarification, the neighbours may collude to promote or demote a neighbouring node as a genuine one or malicious one. Hence robustness of this mechanism is questionable.

In Baburajan and Prajapati [36] have proposed a watchdog mechanism to identify malicious nodes in a wireless sensor networks. Similar to the node detection mechanism proposed in [34], the watchdog mechanism also depends on the broadcast nature of communication in wireless sensor networks. As opposed to the mechanism proposed in [34] where the transmitter itself acts as the monitor listening to the transmission from the intermediate node, in the watchdog mechanism all the nodes who can hear both transmissions can act as the monitors. The identified limitations of the watchdog mechanism include; ambiguous collision, receiver collision, limited transmission power, false misbehaviour and partial dropping. Some of the shortcomings of the watchdog algorithm has been solved by improved algorithms. By creating a cluster head and making it the first level watch dogs can help solve impartial removal, false malicious node, limited power and node conspiracy. Receiver collision problem can be solved by enabling a collision detection mechanism. But this mechanism may not solve the ambiguous collision problem.

Nakul [37] has reviewed several intrusion detection mechanisms that can effectively identify the misbehaving nodes in wireless sensor network. Node misbehaviour in a network may indicate the presence of compromised nodes or malicious nodes introduced by the adversaries or corrupted nodes due to external factors. Irrespective of the reason for misbehaviour, the misbehaving node must be identified and removed from the network. The methods reviewed in [37] include weighted trust evaluation approach, ant colony based approach, data mining based approach, agent based approach, trust based approach, weak hidden Markov model based approach neighbour based approach, game theory based approach and hybrid approach. Details of some of the important approaches are discussed below.

In weighted trust evaluation the sender node assigns trust scores to other nodes in the cluster based on its experience with those nodes. When an intermediate node forwards the frame correctly, its trust score is enhanced. On the other hand, when the forwarded frame does not match the original frame, its trust score is decremented. This algorithm is simple to implement and based on two strong assumptions. They are the base station is honest or not compromised and the majority of nodes in a network are well behaved. This algorithm would fail, if any of these assumption is violated.

In data mining approach applied to the detection of anomalous behaviour of nodes checks all the data packets transmitted in the network. This method has very good detection rate but suffers from the limitation that it requires a lot of processing power and energy to run the data mining algorithms in real time. The main advantages of this method are its ability to detect the anomalous packet before it reaches the access point, to start the detection process immediately without needing any prior training and higher detection rates.

The agent based anomaly detection mechanism employs a combination of both rule based scheme and naive Bayesian technique. This mechanism shows good performance in large distributed sensor networks using common anomaly detection framework with agent learning and distributed data mining techniques.

The trust based approach combines social trust and QoS trust for computing the trust worthiness of a node. Honesty has been used as the parameter for social trust while energy and cooperativeness are the attributes used for computing QoS trust. The final trust score is used for identifying the malicious nodes in the network. The cluster head assigns the trust scores to all the members within the cluster and the cluster heads are similarly evaluated by the base station.

In the weak hidden Markov model based anomaly detection mechanism, state transition probabilities are reduced to rules of reachability. This is a two stage mechanism where in the first stage, the training and learning takes place and in the second stage real time detection of intrusion is carried out. The scoring scheme and deviation detection mechanism introduced as enhancements improves the detection accuracy.

The neighbour based approach exploits the similarity of behaviour in a given community. It is assumed that all the neighbouring nodes in a sensor network would behave similarly due to the fact that they all face similar conditions and limitations. If any node deviates from the common behaviour of its nodes, then it is identified as a malicious node. This approach has better detection rates when the neighbours cooperate with each other with very low false positives and negatives.

The game theory based intrusion detection scheme makes use of a signalling game to model the interaction between nodes in a wireless sensor network. In this mechanism, the interaction between an attacker and a normal has been modelled as a Bayesian game with incomplete information.

In Li et al. [38] have presented survey on methods for detecting node replication attacks in wireless sensor networks. When a sensor node is physically captured by an intruder, it is possible to capture all the information stored within the node. Then he duplicates this node along with inserting his malicious code and then plants them in many strategic locations within the network. The methods presented in [38] include Node to node broadcasting (N2NB), Deterministic Multicast (DM), Randomized Multicast, Randomized, Efficient, Distributed mechanism (RED), Memory Efficient Multicast (MEM), Randomly Directed Exploration (RDE), Distributed detection of node capture attacks, Zone and based Replica Detection, Out of these schemes, some of the important mechanisms are described below.

In node to node broadcasting, every node broadcasts an authenticated message claiming its own location throughout the network. Each node stores the location claim of its neighbours. When a conflict was detected in location claim, the malicious node is revoked immediately. Since the messages from every node in the entire network needs to be processed by every other node, the storage, message and communication cost are high in this scheme. The directed multicast is an improved version of N2NB. In directed multicast, claimer-reporter-witness framework is fully exploited to detect the malicious node efficiently. The claimer shares its location claim to its neighbours and the neighbours act as the reporters. The reporters select a witness using claimer's ID and a function. Then the reporter forwards the claimer's location claim to the witness. If a witness receives multiple claims for the same location, it would then trigger the duplicate node revocation mechanism. This mechanism suffers from one main shortcoming. When an adversary knows the claimer's ID, then it can compute the location of the witness. Hence the adversary can compromise both the claimer and the witness before deploying the malicious node in the network.

Distributed detection of node capture attacks exploits the fact that when a node has been physically captured by an adversary, it will be dormant for a period of time. This protocol measures absence time period of nodes and compares it with a predefined threshold. If the period of absence is more than the threshold value, then it is declared as a compromised node. The effectiveness of this protocol depends on the threshold value.

In Virmani et al. [39] have proposed an exponential trust based mechanism to detect black hole attack in wireless sensor networks. In this mechanism every node maintains a tables and a streak counter in its memory. The table maintains the trust factor of other nodes and the streak counter measures the number of consecutive packets dropped by that node. The trust factor starts with 100 and the streak counter with zero (0) incremented by 1 for every consecutive drop. The streak counter is reset to zero (0) whenever it forwards a packet to the next node. The trust factor for each consecutive drop is computed using the formula $100 * x^i$ where x is a factor less than 1 and i the number of consecutive packets dropped. Since black holes (sinkholes) would be continuously dropping all the packets they receive, their trust value would fall drastically with few packets dropped. This would help identify the sinkholes very fast.

Lim and Choi have proposed malicious node detection mechanism using dual threshold method [40]. In this mechanism two different threshold values are maintained, one for event detection accuracy and the other one for false alarm rate along with trust values for each neighbour in the network. This helps improve the detection of malicious nodes without increasing the overhead.

In Atakli et al. [41] have proposed a weighted trust evaluation to identify malicious nodes by monitoring the reported data. In this work, initially the network is divided into three main groups creating a hierarchical architecture. At the top of the network there are access points or base stations followed by the middle layer occupied by the high powered forwarding nodes. At the lowest level are the low powered sensor nodes with limited functionality. The sensor nodes re organized around high powered forwarding nodes as cluster heads and communicate only with those cluster heads. Only the forwarding nodes have the multi-hop routing capability and assumed to be trustworthy and cannot be compromised. The sensor nodes within the control of a forwarding node are given a weight with 0 and 1 based on its prior behaviour. The forwarding node computes an aggregation results from weighted average of the information received from the sensor nodes within its control. Whenever the reported information of a sensor node deviates from the aggregation results, its trust value is decremented. When the trust value of a given node falls below a pre-decided threshold, it is identified malicious and removed from the network.

In Junior et al. [42] have proposed a malicious node detection scheme through traffic monitoring. In this scheme, all the nodes are considered equal in every sense and communication between the nodes is symmetrical. Every node in the network transmits its node id and the location coordinates obtained from the GPS system. Every node could compute the theoretical received signal power given the identical nature of nodes and the inter-node distance obtained from the location coordinates using the two-ray signal model. When the received signal power is different from the computed theoretical value, the suspicious count maintained in the memory is incremented, otherwise, unsuspicious count would be incremented. When a suspicious message is detected by a node, it transmits the message of suspicion with the id of the suspicious node. Whoever has received this message of suspicion and the original transmission may reply back their opinion based on their own calculations. Then all the nodes within the reach of these nodes updates their opinion (suspicious and unsuspicious) tables based on the opinion received. When the ratio between the suspicious to unsuspicious messages received increases beyond a preset threshold value, it is named malicious and removed from the network.

5 Security of Data in Smart Sensor Networks

Similar to any other network, the data transmitted over a wireless sensor network must also be protected [8]. For any data to be considered as secure, it must satisfy the three security primitives known as confidentiality, integrity and availability. When any of the above security primitive is breached then the security of the data is considered as breached. The confidentiality ensures that only the intended recipient has access to the data and no one else. When data is transmitted over a large network, it may go through many intermediaries before it reaches the final recipient. All the intermediaries must only forward the data towards the recipient but should not be able to read or understand what is in it. Eavesdropping is an attack confidentiality of data. Data integrity assures the recipient that the data received has not been modified or tampered with en-route. In addition to data security, it is also important to ensure source integrity. Source integrity means the data must really be originated by the node where it is claimed to be originated. Impersonation is an attack on source integrity.

Availability is the capability to access the data on a timely fashion, when required to the authorised user. Denial of service is an attack on the availability of data as it prevents the authorised user from accessing the data. This section takes an in depth look at data security in wireless sensor networks. This section takes a detailed look at the prominent work carried out for protecting data in wireless sensor networks.

5.1 Threats to Data in Wireless Sensor Networks

The threats to data collected and transferred in wireless sensor network are not uniform and depends on the type of application [6]. For example, the data collected in a agriculture farm with the aid of a wireless sensor network only requires integrity checking against intentional or unintentional modification. On the other hand, in a military application, the data must be protected for all the three types of security requirements. Namely, confidentiality, integrity and availability [7].

The information transmitted over the wireless channels of a sensor network could be monitored [7]. This is commonly known as eavesdropping. Eavesdropping is a passive attack on the data and can be carried out very easily on a wireless sensor network as the common mode of communication employed in a sensor network is broadcasting. Eavesdropping may not be considered a big issue for many applications such as environmental monitoring or machine health check monitoring in industrial applications. On the other hand, military and medical application require higher security implementation against disclosure of information to unauthorised persons [7]. In military surveillance applications, the information on enemy movements and others of strategic importance are captured and transmitted via a wireless sensor network. If this information falls into the enemy's hand, the consequences would be very serious. Hence it must be protected with the highest level of security available. Similarly, in remote health monitoring applications, the patient information is required to be protected by law. Thus, healthcare applications also require high level security ensuring confidentiality of data. The data confidentiality can be assured by implementing the proper encryption depending on the requirements.

Data injection is another type of attack that can be carried out in a wireless sensor network [7]. Data injection is wrong information introduced into a network by malicious or compromised nodes. Data injection is an active attack where the malicious node actively participates in the network activities. Data injection is more dangerous than eavesdropping as it can affect all types of sensor network applications. Solutions to the issue of data injection is the identification of the malicious node and removing it from the network.

Data modification or corruption is an attack on the integrity of data [7]. Data corruption can happen due to activities of malicious nodes or due to external interferences such as noise. Irrespective of the reason, data corruption must be detected and corrected. Data corruption can be detected using simple hash functions appended to the data or through complicated double encryption techniques [6]. The corrupted packet is generally recovered through retransmission.

Packet deletion in a wireless sensor network may happen due to unintentional dropping of packets as a result of a shortage of resources such as buffer space in a node or a malicious attack such as sinkhole attack and selective packet forwarding [7]. The loss of packets are detected using sequence numbers added to the headers. Unintentionally dropped packets can be recovered through retransmission and if there is a malicious attack on the network, the rogue nodes responsible for the attack must be removed.

Misrouting of packets happen when packet headers get corrupted or due to an active attack on the routing process [7]. If an active attack takes place, the nodes responsible for the attack must be detected and removed. Both packet deletion and misrouting are attacks on the availability of network resources.

5.2 Mechanisms for Protecting Data in Wireless Sensor Networks

Data security in a wireless sensor network is carried out through implementing the right level of encryption of data based on the requirement [6]. For the encryption to be successful, proper distribution and management of keys is a critical requirement [34]. Due to the resource constraints in sensor networks, the conventional key management schemes used in traditional networks cannot be used in wireless sensor networks due to their high overhead and the involvement of external parties [8, 34]. Hence the cryptographic schemes employed in a wireless sensor network must be evaluated to meet the constraints in terms of code size, data size, processing time and power consumption [8].

As public key cryptography has been found to be too expensive to be implemented in a wireless sensor network, many researchers have focussed their attention on secret (symmetric) key cryptography for implementing security in such a constrained environments [8]. When symmetric key cryptography is used, the key management in an open environment becomes a critical issue. In symmetric key cryptographic mechanisms use the same key for encryption as well as decryption. Hence it is essential to transfer the key to the receiver confidentially without the knowledge of the adversary. Also the key management scheme must be capable of handling the addition of new nodes and the removal of existing nodes from the network [8].

Key management schemes can be broadly divided into centralized and distributed key management schemes [8]. In centralized key management, a single node probably the base station carries out all the tasks pertaining to key management including generation, regeneration, distribution and revocation. This single node is known as the key distribution centre. The main shortcomings of this scheme are single point of failure and scalability. On the other hand, in distributed key management schemes, the responsibility of the administration of key is distributed among multiple nodes effectively eliminating the single point of failure and providing better scalability. The distributed key management schemes may use either deterministic or probabilistic distribution algorithms [8].

A key distribution issue can be decomposed into the following steps [34]:

- Key pre-distribution—installing the key in a node prior to deployment.
- Neighbour discovery—discovering the nodes that are just one hop away.
- End to end path key establishment—end to end communication with nodes that are not directly connected.
- Isolating misbehaving nodes—identifying and isolating damaged or malicious nodes.
- Key establishment latency—reducing the latency resulting from communication and power consumption.

Perrig et al. proposed a suit of security protocols for wireless sensor networks in 2002 called SPINS [43]. Within this suite is a secure network encryption protocol (SNEP) that provides confidentiality, integrity and freshness of data through the use of encryption and authentication. The main features of this protocol include the low overhead per message, managing state at every node eliminating the need for transmitting counter values and semantic security. The SNEP also enhances the security of encryption by preceding the data to be encrypted by a random sequence effectively countering the known plain text attack that can be carried out by an attacker. SNEP communicating nodes derive their keys from a shared master key using pseudorandom function. A secure authenticated message using SNEP would be as given in Eq. (2).

$$A \rightarrow B : \{D\}\{K_{AB}, C_A\}, MAC\left(K'_{AB}C_A \parallel \{D\}\{K_{AB}, C_A\}\right) \qquad (2)$$

where A and B are the communicating nodes, D is the data encrypted with derived key K_{AB} and counter value of A; $A; C_A. MAC\left(K'_{AB}C_A \parallel E\right)$ is the message authentication code computed using K'_{AB} the derived key for MAC operation and E the encrypted message.

TinySec is an improved version of SNEP where access control and message integrity are provided through authentication, confidentiality through encryption and semantic security through the use of a unique initialization vector for each invocation of the encryption algorithm [44]. TinySec comes in two specific variants; TinySec-Auth and TinySec-AE. TinySec-Auth provides only authentication using a message authentication code and the payload is left unencrypted. TinySec-AE provides both authentication through the message authentication and encryption of the payload. In TinySec replay protection is not included, hence it must be carried out by a higher layer protocol, if necessary.

Security manager is a method of authenticated key agreement based on public key infrastructure and elliptic cryptography for low rate wireless personal area networks [45]. The security manager gives the static domain parameters such as the base point and elliptic curve coefficients to prospective nodes which use them to establish permanent and ephemeral public keys. Every node in the network computes its own public key and sends it to the security manager which maintains them in its memory. Elliptic curve algorithms provides reasonable computational loads and smaller key sizes for equivalent security compared to RSA the traditional method for public key cryptography. The authenticated key agreement is achieved via security manager based on RC-MQV algorithm that is more advanced than Diffie-Hellman algorithm. RC-MQV is resistant to man in the middle attack, hence security manager is a very robust technique against all known attacks on data in wireless sensor networks as long as the security manager is not attacked and compromised.

In [46], Soroush, Salajegheh and Dimitriou have proposed a strong post deployment key management protocol that is flexible, scalable and robust against node capture attacks. This is a triple key mechanism consisting of pair-wise key, broadcast key and node-base key. Pair-wise key that is established between two neighbours protects their direct one-to-one communication, broadcast key secures the messages sent between neighbours and node-base key protects the communication between a node and the base station. The first step in the operation of this mechanism is the node discovery. Node discovery is carried out through a ping-pong handshake message exchange between neighbours. Once all the neighbours have been discovered, a node will compute its own node-base key and its pair-wise keys and broadcast keys as given in Eq. 3.

$$\left.\begin{aligned} NB_i &= F(i \parallel base\,station\,address \parallel K) \\ PW_{i,j} &= F(min(i,j) \parallel max(i,j) \parallel K) \\ BC_i &= F(i \parallel K) \end{aligned}\right\} \tag{3}$$

where i, $\parallel$, F and K are the node id, concatenation operator, secure pseudo random function and pre-installed global master key respectively.

Secure pseudo random function is implemented using a hash function SHA-1 or MD5. The global master key (K) can be deleted from the memory of nodes later protecting them from falling into enemy hands in times of node capture attacks. Once the calculations are over, node i would be having a complete set of keys for node j. But, node j does not have any information about node i, as it must be sent from i. Node i would then create a message M containing the pair-wise and broadcast keys and encrypt that message with a node-base key derived as follows.

$$NB_j = F(j \parallel base\,station\,address \parallel K)$$

After sending this message, node i will delete the node-base key of node j from its memory leaving only node j capable of decrypting this message. Then global master key K will also be deleted from the memory of node i. The above steps can be followed by a new node when getting added to the network if comes with the pre-installed global master key K. This makes the network resilient to node capture attacks or introduction of malicious nodes by an intruder as he will not have access to K.

The Distributed ANGEL Key Agreement (DAKE) is a direct distributed key establishment based on keying material stored on the nodes [47]. This is an α-Secure Key Establishment process, where α-secure refers to the system that resists the collision up to α entities. Some α-secure keying material KM_{root} stored at a secure location is used to generate an α-secure keying material share KM_i for each entity i in the system. In a typical system, a single symmetric bivariate polynomial $f(x, y)$ of degree α over a finite field $GF(q)$ where q is large enough to accommodate a cryptographic key can be used as KM_{root}. Each entity in the network receives its polynomial share $f(i, y)$ generated by evaluating the original symmetric bivariate polynomial in $x = i$. Two entities in the network (i, j) can agree on a pairwise key by evaluating their respective polynomial shares in the identity of other party as shown in Eq. 4.

$$K_{i,j} = f(i, y)|_{y=j} = f(j, y)|_{y=i} \qquad (4)$$

In DAKE, key segmentation and Horner's rule are used to break the large into multiple sub-polynomials and reduce the number of multiplications by factoring out.

The Modular Architecture for the Security of Sensor Networks (MArSSeNs) is a complete framework of security tools that can provide transparent security individually to all the data streams and network layers of applications in a wireless sensor network [48]. The advantages of MArSSeNs include the implicit and transparent security at any layer of network stack and data stream without requiring changes to application code, elimination code complexity and reduction of errors compared to hard-coded security and facilitation of configuration at both compile as well as run time. MArSSeNs provides in-depth key management and allows distinction between session keys (short term) and encryption keys (long term), using different keys for every service in a data stream policy, controlling maximum key

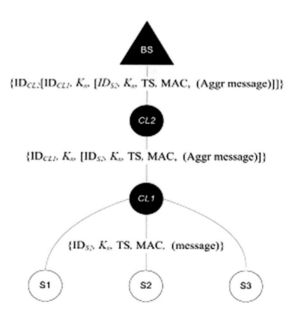

Fig. 2 Key calculation for sensor node S2 up to base station BS [34]

usage and facilities for key establishment, renewal, derivation and revocation. The
MArSSeNs key manager administers the key database, key life cycle and key
management protocols where needed. MArSSeNs supports third party protocols
through a set of interfaces. For non supported key types, it is possible to implement
a sub key manager to handle all the tasks.

The secure triple key management scheme proposed in [34] consists of three
keys. Two of these keys namely the network key and sensor key are pre-installed in
all nodes and the other key is the network generated cluster key addressing the
hierarchical nature of the network. The network key is used for encrypting data and
pass it to the next hop, the sensor key is used by the base station to decrypt and
process it while the cluster leader uses it for decrypting and passing the data to the
base station and the cluster key is used for decrypting data and passing it to the
cluster leader. Figure 2 shows the key calculation process at different levels.

6   Routing Security in Smart Sensor Networks

Wireless sensor network is a infrastructureless multi-hop network, where the transfer
of data takes place by forward from one node to another. In such a network, routing
plays an important role in carrying the information from the source to destination
[49]. The routing protocols employed determines the best route to transfer the data
from the source to destination possibly the base station.

The routing protocols in wireless sensor networks can be grouped into three main
categories according to the network structures [8]. There are namely (i) flat-based
routing, (ii) hierarchical-based routing and (iii) location-based routing. In flat-based
routing, all nodes are considered equal and assigned similar roles, in hierarchical
routing, nodes are assigned different role and in location based routing, the geographical location of the nodes are used for routing data in the network.

Many parameters of a wireless sensor network including the end-to-end delay,
packet delivery ratio, life time of the network etc., depends on the performance of
the routing protocol [49]. Due to the nature of wireless sensor networks, the routing
protocols may employ different criteria for selecting best path such as low energy
consumption path or least hostile path compared to traditional network where the
shortest path or the least congested path is generally selected as the best path [50].

The routing protocols in a wireless sensor network is also vulnerable to several
attacks [49]. The attacks carried out on routing protocols have severe consequences
due to the self contained and self configuring nature of the network itself. In order
to overcome these threats, several secure routing protocols have been proposed in
the literature [49]. This section takes an in depth look at the security threats for
routing and the mechanism proposed for overcoming them.

6.1   Threats that Affect Routing in Wireless Sensor Networks

There are several attacks that directly target the routing protocols for disrupting the traffic in a network [8]. The attacks on a routing protocol may create routing loops, attracting or repelling traffic from a selected set of nodes, extend or shorten source routes, generate fake error messages, partition the network or extend the end to end delay. Some of the most common attacks on the routing process are described below.

Spoofed routing information: This is direct attack against a routing protocol targeting the routing information. Routing protocols require exchanging routing information between nodes for building a routing table with the most current status of the network. The routing table must be up-to-date with the status of nodes as this information is used by nodes to identify the best path to the destination. An attacker may spoof, alter or replay routing information effectively disrupting the traffic flow in a network. This attack may lead to many problems such as routing loops, increased end to end latency and even network partitioning [50].

Sinkhole attack: In this attack, a malicious node has been shown as the most attractive next hop node to forward the packet towards the destination. Once a packet reaches the malicious node, it is dropped instead being forwarded.

Sybil attack: A single node presents itself with multiple identities which are either stolen ones or fabricated ones. When a Sybil attack has been carried on routing, it makes multiple routes to go through a single compromised node effectively delaying or dropping packets en-route.

HELLO flood: Many routing protocols assume that the HELLO messages come only from a neighbouring node. A malicious node with a high powered transmitter may fool many nodes as it is within their neighbourhood effectively announcing a false shorter route to the base station. All the nodes receiving this HELLO message would try to forward the packets to this malicious node though it is outside their range.

Acknowledgement spoofing: Some routing algorithms require the transmission of acknowledgement messages for proper operation. A malicious node eavesdropping on the conversation of other nodes may spoof their acknowledgement packets. This disseminate wrong information about nodes.

6.2   Secure Routing Protocols

In order to overcome the threats and attacks on routing, many researchers have proposed secure routing protocols that can withstand these attacks. Many of these protocols make use of cryptographic primitives and authentication mechanisms to minimize the effects of attacks, while others make use of trust between nodes identify the malicious or compromised nodes.

In Duan et al. [49] have proposed a lightweight and secure routing scheme. This scheme makes use of trust computed between nodes to identify the best path to the destination. The routing algorithm and the operation of the scheme is as follows:

Step 1: When the node v_0 wants to send a packet to v_{11}, which is not its neighbour, it sends a trust request packet to its neighbours. A trust request is a 6-ary tuple and is denoted by $TR = s_{id}, t_{id}, t(p)_{th}, ts, s, hl$, where $s_{id}, t_{id}, t(p)_{th}, ts, s$ and hl are source and destination node ids, threshold of path trust, timestamp, sequence number and hop limit of trust request packet respectively.

Step 2: A neighbouring node receiving this request will check, if the destination node v_{11} is in its neighbour list. if yes, it replies to the request with the trust value of the destination hop, else it broadcasts the requests to all its neighbours. All the neighbours who initiated the process would process this request.

Step3: This process continues until the request reaches a node in whose neighbour list the destination is found. Then the reverse process initiated through the selected path (through which the request came) with the trust value of path until the original requester node v_0.

Step 4: The originator evaluates the paths received, if more than one is received and selects the path with the highest trust.

Step 5: v_0 forwards the data packet through the selected path.

The routing algorithm and operation of the scheme are shown in Figs. 3 and 4 respectively.

The lightweight secure routing scheme may not be as secure as it has been claimed to be and come under many attacks when there are malicious nodes in the networks. The best path selected purely depends on the trust scores transmitted by intermediate nodes. This can be exploited by the adversary to mislead the requester to select non optimal paths, worse sometimes towards sinkholes. If the malicious nodes collude, the effect would be worse.

Ambient Trust Sensor Routing (ATSR) proposed by Zahariadis et al. [51] follows the geographical approach. The main criteria of the next hop selection in this mechanism is the geographical coordinates along with the remaining energy and trust value of the node. The combination of the multiple input parameters make the protocol more rugged and help lengthen the life of the network by not exploiting the best (having highest score) node as it might drain their battery very soon. The trust computation process takes many criteria including packet forwarding efficiency, network layer acknowledgements, message integrity, node authentication, confidentiality, reputation response and reputation validation as inputs making it a very comprehensive process and less vulnerable to attacks by an intruder that provides false information. The energy computation mechanism is the weakest link in the process. Since the remaining energy is expressed as a percentage of original energy, if all the nodes are not of the same capacity, this information may mislead the nodes to select a node with lower level of absolute energy when better nodes with large energy levels are present in the network.

Secure routing mechanism proposed in [34] uses only the cluster heads to forward the encrypted data towards the base station. The routing protocol is divided

(1) Process Initialization
(2) $G_R^*(V, E_R, r) = v_n$, v_n is the destination node
(3) Add v_n to V^*, V^* represents the set of nodes that have optimal routes to v_n
(4) **while** $V \neq V^*$ **do**
(5)   **for all** node $v_i \in V - V^*$ **do**
(6)   Sort $r(p(v_i, v_n))$
(7)   Obtain $\vec{r}(p(v_i, v_n)) \triangleq (q_0, q_1, \ldots, q_m)$
(8)   where $q_0 = t(p(v_i, v_n))$
(9)   **for all** $v_k \in \Gamma(v_i)$ **do**
(10)   **if** $t(v_i, v_k) \otimes_T t(p(v_k, v_n)) \succeq t(p(v_i, v_n))_{\text{th}}$ **then**
(11)   Add $(v_i, p(v_k, v_n))$ to $P_{Q_0}^*(v_i, v_n)$
(12)   **end if**
(13)   **end for**
(14)   **if** $P_{Q_0}^*(v_i, v_n) = \emptyset$ **then**
(15) v_i is disconnected from the network
(16)   Continue;
(17)   **end if**
(18)   **for** $j = 1; j < m; j ++$ **do**
(19) $P_{Q_j}^*(v_i, v_n) = \oplus_{Q_j} P_{Q_{j-1}}^*(v_i, v_n)$
(20)   where $p_{Q_{j-1}}^*(v_i, v_n) \subseteq P_{Q_{j-1}}^*(v_i, v_n)$
(21)   **end for**
(22)   **if** $P_{Q_m}^*(v_i, v_n) = \emptyset$ **then**
(23) v_i is disconnected from the network
(24)   Continue;
(25)   **else**
(26)   Add v_i to V^*
(27)   Add $P_{Q_m}^*(v_i, v_n)$ to $G_R^*(V, E_R, r)$
(28)   **return** $p_R^*(v_i, v_n), p_R^*(v_i, v_n) \subseteq P_{Q_m}^*(v_i, v_n)$
(29)   **end if**
(30)   **end for**
(31) **end while**
(32) END Process

Fig. 3 Lightweight secure routing algorithm

into two categories; one for sending data from a sensor node to the base station and the one for sending information from the base station to the sensor nodes.

The algorithm for sending data from a sensor node to the base station is as follows:

Step 1: Request the cluster key K_c from cluster leader.
Step 2: Use K_c and its own key K_n to compute the encryption key K_{cn}.
Step 3: Encrypt the data with K_{cn} and append its node ID and current time stamp TS and forward the packet to the cluster head.
Step 4: The cluster head upon receiving the encrypted data packet, appends its ID and forwards it to the base station, if directly connected, otherwise forwards it to another cluster head.

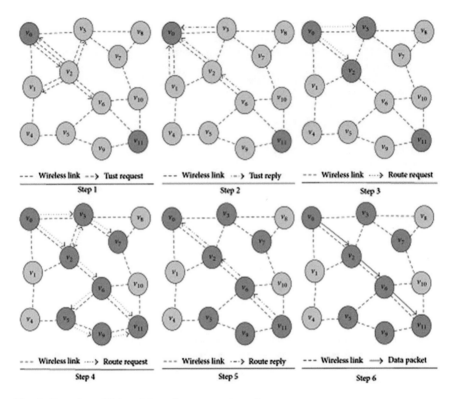

Fig. 4 Operation of lightweight and secure routing scheme

Figure 5 shows the node algorithm in detail.

When the base station wants to broadcast any data to sensor nodes, it just encrypts the data packet with sensor key K_s and forwards it to the directly connected cluster heads.

In this scheme, the cluster heads are assumed to be non-compromisable, this may not be 100 % correct. When a cluster head is compromised, the entire security of the system may fail.

Intrusion-tolerant routing mechanism in wireless sensor networks (INSENS) proposed in [52] builds routing tables in each node bypassing the malicious nodes in the network. Control information pertaining to routing is authenticated by the base station for the purpose of preventing injection of false routing data. The base station computes and disseminates the routing tables to all the nodes helping the nodes saving their energy. Redundant multi-path routing enables the nodes to overcome the sinkhole and wormhole attacks carried out by malicious nodes.

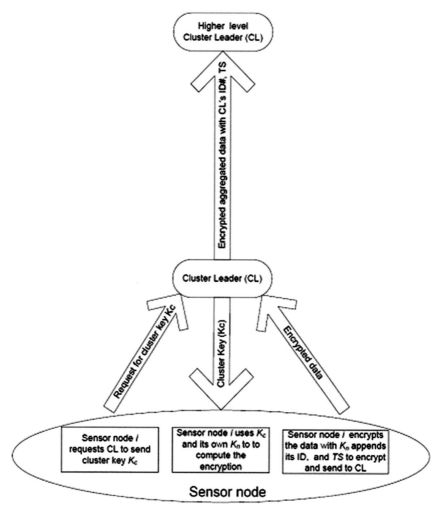

Fig. 5 Node routing algorithm [34]

The route discovery mechanism of INSENS is as follows:

Step 1: Base station sends a request message to all nodes through multi-hop forwarding.

Step 2: Nodes receiving the request message, records the identity of the sender and forwards it to their neighbours for the first time (repeated flooding not allowed).

Step 3: Nodes respond with their local topology by sending feedback messages.

Step 4: Base station calculates forwarding tables for all nodes with two independent paths for each node and disseminates them.

In this communication, the integrity of all the messages including requests and feedbacks are protected using encryption by a shared key mechanism. In this protocol, when a malicious node does not forward the message, it can reach the destination through another path. Hence the effect of the sinkhole attack is minimised, if not eliminated. The malicious nodes may also send spurious messages to drain the battery power in the downstream nodes.

Trust routing for location aware sensor networks (TRANS) proposed in [53] is a location aware routing protocol. TRANS makes use of a loose-time synchronization asymmetric cryptographic scheme to ensure message confidentiality. The operation of the protocol is as follows:

Step 1: The base station broadcasts an encrypted message to all its neighbours.
Step 2: Neighbours receiving this message, decrypt it add their locations encrypt and forwards it to its neighbours closer to the destination.

The security of this protocol is ensured by encryption. Only the trusted nodes can decrypt the messages as only they possess the shared key. The destination node authenticate the received message using the message authentication code added by the base station.

The acknowledgements and replies from the sensor nodes to the base station just traverse the reverse path through which the message arrived.

The secure route discovery protocol proposed in [54] guarantees correct topology discovery in an ad hoc sensor networks. This protocol ensures security of messages through message authentication code and accumulation of node identities along the route traversed by the message. Each node in the network discovers every other node using the node identities appended to the messages finally discovering the entire network topology. The verification of the message authentication protocol at both source and destination ensures the integrity of the messages.

The ant colony-based routing protocol proposed in [55] consists of four distinct stages in setting up a secure route to destination. In stage 1, clusters are formed based on their geographical regions. Within each region a node N and a parameter L are chosen randomly where L indicates the level of neighbours in the cluster. Using limited HELLO floods, the neighbour list exchange process starts from node N to L levels. In stage 2, cluster heads are chosen. Within each cluster formed, three nodes H_1, H_2 and H_3 are chosen randomly and their resource levels are computed. The node with the highest resource level is selected as the cluster head. In stage 3, the routing process starts. The node with data to be sent forwards its message to the cluster head. Then the cluster head sends HELLO messages along with pheromone request to its neighbouring cluster heads. The entire neighbour cluster heads reply to the request with their current pheromone values. This process is repeated until a optimum path is found to the destination. The elimination of malicious nodes in this protocol is achieved through conformity checks carried out at the end of cluster formations.

7 Future Directions in Smart Sensor Network Security

Though extensive work has been carried out in various aspects of wireless sensor networks security, there are still many open problems that need to be addressed. This section takes a brief look at the some of the open areas.

Currently security in wireless sensor network research is carried out in a fragmented manner each group concentrating on specific problems and aspects. It is necessary to have a more unified approach towards various aspects of the security in wireless sensor networks. Hence it is necessary to produce a uniform application independent security framework for wireless sensor networks.

Generally the implementation and enhancement of security affects the other aspects of sensor networks such as user friendliness and quality of service. This would normally affect the usefulness and usability of these networks. It is necessary to have security implementations that have minimal impacts on other aspects of wireless sensor networks.

hough some research has already been carried out and obtained some promising results on the use of public key cryptography in wireless sensor networks, it is still an open area. The code size, processing time and power consumption are still high for the deployment of them widely. Hence an active look into this area would be a worthwhile effort. The specific areas that can be looked at include code optimization, energy efficient computation, and optimization of private key operations.

Wireless sensor nodes are deployed in an open area that is not only harsh but also hostile. Hence the sensor nodes face several threats from natural as well as manmade sources. Hence the security of the sensor nodes must be increased. The improvement of sensor node security requires a multi-pronged approach including physical, logical and technological aspects.

In wireless sensor network secure routing arena, the following areas need further investigations.

– Energy optimized routing protocols: In any network, though routing is an essential requirement, the operation of routing protocols is an overhead. Hence the overhead incurred in the operation of routing protocols must be reduced as much as possible.
– Faster convergence: The scale of operation of wireless sensor networks is large with thousands of nodes. Also the topology is also dynamic compared to conventional networks. Under these circumstances, the routing table would also constantly undergo rapid changes. Thus routing protocols with faster convergence times is an immediate requirement.
– The routing protocols and information face attacks by various threats and these would increase in the future with the popularity of wireless sensor networks. Hence it is necessary to have more secure routing protocols that are robust and resilient in the face of increased attacks in the future.

8 Conclusions

Sensor nodes have become more intelligent in recent times due to the developments in many fields including VLSI design, computing and communication. With the increased intelligence incorporated into the sensor nodes, the application areas where these nodes can be used has also increased. Along with the increased popularity and deployments of wireless sensor networks, the threats and attacks on these networks have also become a major issue demanding immediate attention to them. Several research groups are working on enhancing the security of these networks and proposed many mechanisms, techniques and algorithms. This chapter took an in depth look at the security implementations in wireless smart sensor networks from three specific angles; namely sensor node security, data security and routing security. Though tremendous work has already been done in the area of wireless sensor network security, still there is a lot room for future work in this area.

References

1. Mohamed, M.I., Wu, W.Y., Moniri, M.: Power harvesting for smart sensor networks in monitoring water distribution system. IEEE International Conference on Networking, Sensing and Control, pp. 393–398, Delft, The Netherlands (2011)
2. Zhang, Y., Gu, Y., Vlatkovic, V., et al.: Progress of smart sensor and smart sensor networks. 5th World Congress on Intelligent Control and Automation, pp. 3600–3606, Hangzhou, China (2004)
3. Lyle, A.C., Naish, M.D.: A software architecture for adaptive modular sensing systems. Sensors 10(8), 7514–7560 (2010)
4. Kizza, J.M.: Implementing security in wireless sensor networks. 4th Annual International Conference on Computing and ICT Research, pp. 296–311, Kampala, Uganda (2008)
5. Razzak, M.I., Elmogy, B.A., Khan, M.K., et al.: Efficient distributed face recognition in wireless sensor network. Int. J. Innovative Comput. Inf. Control 8(4), 2811–2822 (2012)
6. Sharma, K., Ghose, M.K., Kuldeep, : Complete security framework for wireless sensor networks. Int. J. Comput. Sci. Inf. Secur. 3(1), 1–7 (2009)
7. Boyle, D., Newe, T.: Securing wireless sensor networks security architectures. J. Netw. 3(1), 65–77 (2008)
8. Sen, J.: A survey on wireless sensor network security. Int. J. Commun. Netw. Inf. Secur. 1(2), 55–78 (2009)
9. Hecker, M., Karol, A., Stanton, C., et al.: Smart sensor networks: communication, collaboration and business decision making in distributed complex environments. International Conference on Mobile Business, pp. 242–248, Sydney, Australia (2005)
10. Herrera-Quintero, L.F., Macia-Perez, F., Ramos-Morillo, H., et al.: Wireless smart sensors networks, systems, trends and its impact in environmental monitoring. IEEE Latin-American Conference on Communications, pp. 1–6, Medellin, Colombia (2009)
11. Fine, G.F., Cavanagh, L.M., Afonja, A., et al.: Metal oxide semi conductor gas sensors in environmental monitoring. Sensors 10(6), 5469–5502 (2010)
12. Hancke, G.P., Silva, B.C., Hancke, G.P.: The role of advanced sensing in smart cities. Sensors (Basel) 13(1), 393–425 (2013)
13. Fan, G., Wang, R., Huang, H., et al.: Coverage-guaranteed sensor node deployment strategies for wireless sensor networks. Sensors (Basel) 10(3), 2064–2087 (2010)

14. Singh, A., Sharma, T.P.: A survey on area coverage in wireless sensor networks. International Conference on Control, Instrumentation, Communication and Computational Technologies, pp. 900–907, Kumaracoil, Thuckalay, TN, India (2014)
15. Taniguchi, Y., Kitani, T., Leibnitz, K.: A uniform airdrop deployment method for large-scale wireless sensor networks. Int. J. Sens. Netw. **9**, 182–191 (2011)
16. Filippou, A., Karras, D.A., Papademetriou, R.C.: Coverage problem for sensor networks: an overview of solution strategies. 17th Telecommunications Forum, pp. 134–136, Serbia, Belgrade (2009)
17. Sheikhpour, R., Jabbehdari, S., Khadem-Zadeh, A.: Comparison of Energy efficient clustering protocols in heterogeneous wireless sensor networks. Int. J. Adv. Sci. Technol. **36**, 27–40 (2014)
18. Ranjan, R., Kar, S.: A novel approach for finding optimal number of cluster head in wireless sensor network. National Communications Conference, pp. 1–5, Bangalore, India (2011)
19. Ghosh, A., Das, K.S.: Coverage and connectivity issues in wireless sensor networks. In: Shorey, R., Ananda, A.L., Chan, M.C., et al. (eds.) Mobile, wireless and sensor networks: Technology, applications, and future directions. Wiley, New York (2006)
20. Li, J., Andrew, L.L.H., Foh, C.H., Zukerman, M., et al.: Connectivity, coverage and placement in wireless sensor networks. Sensors **9**, 7664–7693 (2009)
21. Khelifa, B., Haffaf, H., Madjid, M., et al.: Monitoring connectivity in wireless sensor networks. Int. J. Future Gener. Commun. Networking **2**(2), 1–10 (2009)
22. Zhang, H., Hou, J.C.: Maintaining sensing coverage and connectivity in large sensor networks. Ad Hoc Sens. Wireless Netw. **1**, 89–124 (2005)
23. McDermott-Wells, P.: What is bluetooth? IEEE Potentials **23**(5), 33–35 (2005)
24. Ting, K.S., Ee, G.K., Ng, C.K., et al.: The performance evaluation of IEEE 802.11 against IEEE 802.15.4 with low transmission power. 17th Asia-Pacific Conference on Communications, pp. 850–855, Kota Kinabalu, Sabah, Malaysia (2011)
25. Liu, T., Liu, J., Liu, B.: Design of intelligent warehouse measure and control system based on Zigbee WSN. International Conference on Mechatronics and Automation, pp. 888–893, Xi'an, China (2010)
26. Teng, Z., Kim, K.I.: A survey on real-time MAC protocols in wireless sensor networks. Commun. Netw. **2**(2), 104–112 (2010)
27. Carman, D.W., Krus, P.S., Matt, B.J.: Constraints and approaches for distributed sensor network security. Technical Report 00-010, Network Associates Inc., Glenwood, MD, USA (2000)
28. Hill, J., Szewezyk, R., Woo, A., et al.: System architecture directions for networked sensors. 9th International Conference on Architectural Support for Programming Languages and Operating systems, pp. 93–104, Cambridge, MA, USA (2000)
29. Tsiftes, N., Dunkels, A., He, Z., et al.: Enabling large-scale storage in sensor networks with the coffee file system. International Conference on Information Processing in Sensor Networks, pp. 349–360, San Francisco, CA, USA (2009)
30. Marigowda, C.K., Shingadi, M.: security vulnerability issues in wireless sensor networks: A short survey. Int. J. Adv. Res. Comput. Commun. Eng. **2**, 2765–2770 (2013)
31. Kumar, Y., Munjal, R., Kumar, K.: Wireless sensor networks and security challenges. Int. J. Comput. Appl. RTMC **9**, 17–21 (2012)
32. Perrig, A., Stankovic, J., Wagner, D.: Security in wireless sensor networks. Commun. ACM **47**(6), 53–57 (2004)
33. Raman, B., Chebrolu, K.: Censor networks: A critique of sensor networks from a systems perspective. ACM SIGCOMM Comput. Commun. Rev. **38**(3), 75–78 (2008)
34. Zia, T.A., Zomaya, A.Y.: A lightweight security framework for wireless sensor networks. J. Wireless Mobile Netw., Ubiquitous Comput. Dependable Appl. **2**(3), 53–73 (2011)
35. Tsitsigkos, A., Entezami, F., Ramrekha, T.A., et al.: A case study of internet of things based on wireless sensor networks and smart phones. 28th Wireless World Research Forum Meeting, pp. 1–10, Athens, Greece (2012)

36. Baburajan, J., Prajapati, J.: A review paper on watchdog mechanism in wireless sensor network to eliminate false malicious node detection. Int. J. Res. Eng. Technol. **3**(1), 381–384 (2014)
37. Nakul, P.: A survey on malicious node detection in wireless sensor networks. Int. J. Sci. Res. **2** (1), 691–694 (2013)
38. Li, W.T., Feng, T.H., Hwang, M.S.: Distributed detecting node replication attacks in wireless sensor networks: a survey. Int. J. Netw. Secur. **16**(5), 323–330 (2014)
39. Virmani, D., Hemrajani, M., Chandel, S.: Exponential trust based mechanism to detect black hole attack in wireless sensor network. Int. J. Soft Comput. Eng. **4**(1), 14–16 (2014)
40. Lim, S.Y., Choi, Y.H.: Malicious node detection using dual threshold in wireless sensor networks. J. Sens. Actuator Netw. **2**, 70–84 (2013)
41. Atakli, I.M., Hu, H., Chen, Y., et al.: Malicious node detection in wireless sensor networks using weighted trust evaluation. Symposium on Simulation of System Security, pp. 836–843, Ottawa, Canada (2008)
42. Junior, W.R.P., Figueiredo, T.H.P., Wong, H.C., et al.: Malicious node detection in wireless sensor networks. 18th International Parallel and Distributed Processing Symposium, Santa Fe, NM, USA (2004)
43. Perrig, A., Szewczyk, R., Tygar, J.D., et al.: SPINS: security protocols for sensor networks. Wireless Netw. **8**(5), 521–534 (2002)
44. Karlof, C., Sastry, N., Wagner, D.: TinySec: a link layer security architecture for wireless sensor networks. 2nd International Conference on Embedded Networked Sensor Systems, pp. 162–175, Baltimore MD, USA (2004)
45. Heo, J., Hong, C.S.: Efficient and authenticated key agreement mechanism in low rate WPAN environment. International Symposium on Wireless Pervasive Computing, pp. 1–5, Phuket, Thailand (2006)
46. Soroush, H., Salajegheh, M., Dimitriou, T.: Providing transparent security services to sensor networks. IEEE International Conference on Communication, pp. 3431–3436, Glasgow, Scotland (2007)
47. Garcia-Morchon, O., Baldus, H.: The ANGEL WSN Security Architecture. Third International Conference on Sensor Technologies and Applications, pp. 430–435, Athens, Greece (2009)
48. Cionca, V., Newe, T., Dadarlat, V.: MArSSeNs: a modular architecture for the security of sensor networks. IEEE Sensors, pp. 1209–1212, Limerick, Ireland (2011)
49. Duan, J., Yang, D., Zhu, H., et al.: TSRF: a trust aware secure routing framework in wireless sensor networks. Int. J. Distrib. Sens. Netw. 1–14 (2014)
50. Xiangyu, J., Chao, W.: The security routing research for WSN in the application of intelligent transport system. IEEE International Conference on Mechatronics and Automation, pp. 2318–2323, Luoyang, Henan, China (2006)
51. Zahariadis, T., Leligou, H.C., Voliotis, S., et al.: Energy-aware secure routing for large wireless sensor networks. WSEAS Trans. on Commun. **9**(8), 981–991 (2009)
52. Deng, J., Han, R., Mishra, S.: INSENS: intrusion-tolerant routing in wireless sensor networks. Technical report CU-CS-939-02, Department of Computer Science, University of Colorado, Boulder, CO, USA (2002)
53. Tanachaiwiwat, S., Dave, P., Bhindwale, R.: Routing on trust and isolating compromised sensors in location-aware sensor networks. 1st International Conference on Embedded Networked Sensor Systems, pp. 324–325, Los Angeles, CA, USA (2003)
54. Papadimitratos, P., Haas, Z.J.: Secure routing for mobile ad hoc networks. SCS Communication Networks and Distributed Systems Modeling and Simulation Conference, pp. 1–13, San Antonio, TX, USA (2002)
55. Arolkar, H.A., Sheth, S.P., Tamhane, V.P.: Ant colony based approach for intrusion detection on cluster heads in wireless sensor networks. International Conference on Communication, Computing and Security, pp. 523–526, Rourkela, Odisha, India (2011)

Author Biography

Mohamed Fazil Mohamed Firdhous   Mohamed Fazil Mohamed Firdhous is a Senior Lecturer and the Director of Postgraduate Studies at the Faculty of Information Technology, University of Moratwua, Sri Lanka. He is engaged in undergraduate and postgraduate teaching along with cutting edge research in the areas of trust and trust management for cloud computing, Internet of Things, mobile adhoc networks, vehicular networks, computer security and rural ICT development. He has teaching, research and industry experience in many countries including Sri Lanka, Singapore, United States of America and Malaysia. In addition to his teaching and research activities at the University, he is a highly sought after ICT consultant to the government and private institutions in Sri Lanka.

Automatic Music Composition from a Self-learning Algorithm

Michele Della Ventura

Abstract There are several automatic music composition algorithms that generate a classical music melody or Jazz and Blues chords and progressions. These algorithms often work by applying a series of music rules which are explicitly provided in order to determine the sequence of the exit codes. In other cases, the algorithms use generation techniques based on Markov Models. The objective of this article is to present an algorithm for the automatic composition of classical tonal music, based on a self-learning model that combines De La Motte's theory of Functional Harmony in a Markov process. This approach has the advantage of being more general compared to the explicit specification of rules. The article is going to demonstrate the effectiveness of the method by means of some examples of its production and is going to indicate ways to improve the method.

Keywords Artificial intelligence · Automatic composition · Harmonic function · Hidden Markov model · Music · Self-learning

1 Introduction

One of the main objectives in the field of artificial intelligence (AI) is to develop systems able to reproduce intelligence and human behavior: the machine is not expected to be able to have the same cognitive abilities as humans, or to be aware of what it is doing, but only to know how to efficiently and optimally solve problems, being them difficult ones, in specific fields of action. Therefore, the purpose of the studies carried out in the field of AI is not to replace human beings in all their capacities, but to support and improve human intelligence in certain specific fields: the improvement may be based on the computing power derived from the use of computers.

M.D. Ventura (✉)
Department of Technology, Music Academy "Studio Musica", Treviso, Italy
e-mail: dellaventura.michele@tin.it

© Springer International Publishing Switzerland 2016   223
K.K. Ravulakollu et al. (eds.), *Trends in Ambient Intelligent Systems*,
Studies in Computational Intelligence 633, DOI 10.1007/978-3-319-30184-6_9

The areas in which the studies on AI have been developed are generally the areas of multi-agent systems, automatic learning, natural language processing, planning, robotics and vision, web.

One of the research fields which is still partially involved in the AI-related process is music.

The themes of interest of this field refer mainly to the recovery of paper music scores, the recovery and preservation of audio media, the study and realization of databases for music, of protection models for the cultural musical patrimony, of models for the distribution and fruition of music and of models for the segmentation of the score.

There have yet only been a few attempts in the field of musical composition: to compose music is an art and it is a difficult task even for human beings. When a composer writes a musical piece, he has an idea, an intention and has his creativity. A musical piece is a multi-dimensional space with different interdependent levels: duration of sounds, musical phrases, vertical and horizontal sonorities, dynamics, articulations, and so on.

To automatize the task of music composition turns out to be rather difficult, if not impossible.

This article is going to present an algorithm able to generate a musical idea, of assistance to the composer, on the basis of a self-learning system, centered on the concept of "Functional Harmony".

This paper is structured as follows. We start by reviewing background and related work in Sect. 2. The theory of the Functional Harmony is described in Sect. 3. The Process of Markov is described in Sect. 4. We discuss the methods and initial results in Sect. 5. Section 6 contains the conclusions.

2 Background and Related Work

As opposed to other areas, music is a research area which has been yet little explored as far as AI is concerned.

Several studies have been carried out on computer-aided musical analysis or the processing of already-written music texts, yet there have been only a few studies in the field of automatic composition.

A first interesting attempt is the work of Cambouropoulos [1], that, starting from the concept of causality, uses Markov's Chains (a principle that will be used in consequent studies [2, 3]) as a tool to help the generation of a musical idea.

Another interesting analysis is the work of D. Cope: *Experiments in Musical Intelligence* (EMI) [4]. It is corpus-driven and adopts techniques of pattern matching, musical recombinancy, and augmented transition networks, a technique commonly used in natural language processing.

The concept lying at the basis of this study is that of recombination of the musical phrases found in already existing music compositions. The result is that not only the produced music is pleasant, but it also tries to produce music that copies the style of a composer.

Following Cope's work logic, a different system known as the *Automated Composer of Style-Sensitive Music* (ACSSM) [5] is developed.

Like EMI, the output music is produced by reconstructing the deconstructed music segments. Compared to EMI, the techniques used for deconstruction and reconstruction of ACSSM are improved by adopting structures proposed in a preference rule based theory called *A Generative Theory of Tonal Music* [6], which models the unconscious intuitions perceived by music listeners.

Specifically, the lengths of segments depend on the *grouping structure* of the original music; this technique turns out to perform better than the discretely sizing technique used in EMI. ACSSM also considers the metrical aspect of music by attempting to imitate the *metrical structure* of the original music in the output music.

Other music generators include Vox Populi [5], which is an interactive system for music composition, based on genetic algorithms, and *Band-OUT-of-a-Box* [7, 8], which is an interactive real-time improviser based on several machine learning techniques, including clustering and Markov chains.

This article presents an algorithm, inspired by preceding works, that has the objective of generating a new "musical idea", i.e., a sequence of notes which, as a whole, form an idea for the composer.

Though simple, the new musical idea must stem from a well-defined compositional logic that is not formalized beforehand, but that is going to be automatically and gradually built, by analyzing the harmonic structure contained in the already existing musical compositions (of tonal style and by different authors).

The algorithm created for this purpose will have the main task of reading not the simple harmonic structure characterizing every single movement from different musical compositions, but the "harmonic function" that is carried by every single movement, as specified by De la Motte's *Theory of the functional harmony*.

Using the Markov process, the algorithm will be able to improve ever more the quality of the musical ideas, by reading ever more musical compositions.

An interesting aspect of our system, already emerged in EMI, is that the new "musical idea" should, not only be appreciable, but, when reading compositions of the same author, copy the style of the same author.

3 Functional Harmony

In the functional theory [9], the goal is to identify in a sound, a chord or a chord succession, the "intrinsic sonorous value" assumed, compared to a specific reference system polarized in a center, or the capacity to establish organic relations with other sounds, chords or chord successions of the same system.

The functional theory tends to go beyond the sonorous event as it manifests itself, to interpret what lies behind that which appears in a particular instant, to seize the meaning, the "role" that it covers in comparison to other events that come before and after it, therefore the "function" that it performs in the context within which it is immersed.

In particular, as far as the chord is concerned, the functional theory tends to research, beyond what it represents by itself in comparison to a certain reference system [10] (for instance, the chord G-B-D, compared to the tonal system and the tonality of C Major, is the dominant chord), the harmonic function performed, the organic relation established with the one that comes before and the one that comes after it.

The pillars of the functional theory are the harmonic functions of tonic (T), sub-dominant (S) and dominant (D), that Riemann was the first to identify as the foundation and pivot of any type of chord succession, hypothesizing in the connection I-IV-V-I (Fig. 1) the archetype of the tonal harmony and the model which any type of chord concatenation should be traced back to (Fig. 2).

It follows that all the chords will have a harmonic function of relaxation or of tonal center T, or of tension towards such center D, or of breakaway from it S.

The three harmonic functions of I, IV and V degree are termed main functions because they are linked by a relation based on the interval of the perfect 5th that separates the keynotes of the three corresponding chords; the chords relating to the rest of degrees on the scale (II, III, VI and VII) are considered "representatives" of the I, IV and V degree (with which there is an affinity of the third—two sounds in common—because the 3rd is actually the interval that regulates the distance between the respective keynotes) and secondary harmonic functions rest with it.

Fig. 1 Archetype of the tonal harmony according to Riemann

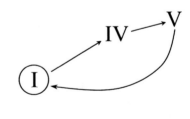

Fig. 2 Riemann's analysis of Beethoven's Piano Sonata n° 1 op. 2

Fig. 3 Harmonic functions of
the degrees of the scale

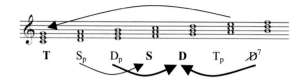

Figure 3 illustrates the sequence of the degrees of the C major scale with their related harmonic function, in which it is easy to notice the functional correlation among the different degrees.

Based on the above considerations it is possible to infer how a musical phrase is built on the basis of the sounds belonging to the chords of certain degrees of the scale that follow one another according to the diagram in Fig. 1. To this end, it is important to point out that generally:

1. the function of Tonic (T) goes towards a function of Subdominant (S) that can be represented by the IV degree (S) or by the II degree (S$_p$) of the scale;
2. the function of Subdominant (S) goes towards a function of Dominant (D) that can be represented by the V degree (D), by the III degree (D$_p$) or by the VII degree (D^7);
3 the function of Dominant (D) goes towards a function of Tonic (T) that can be represented by the I degree (T) or by the VI degree (T$_p$).

The adverb "generally" was used to describe the direction of the tonal functions because the composer has a certain degree of freedom of writing that in some cases allows him to disregard the provisions of musical grammar: for instance, the function of Tonic (T) might go directly towards the function of Dominant (D).

The algorithm developed on the basis of the theory of Functional Harmony represents, therefore, a support to the composer: it can create and propose a new musical idea that the composer may modify by enriching it with the melodic figurations.

Musical grammar provides the composer with a series of tools allowing him to vary, within the same musical piece, an already presented melodic line, by inserting notes which are extraneous to harmony. The sounds of a melodic line, in fact, may belong to the harmonic construction or may be extraneous to it. The former sounds, which fall in the chordal components, are called real, while the latter sounds, which belong to the horizontal dimension, take the name of melodic figurations (passing tones, turns or escape tones).

They are complementary additional elements of the basic melodic material that lean directly or indirectly on real notes and also resolve on them. The use of melodic figurations, therefore, allows achieving greater freedom of the melody, bestowing upon it a better profile (see Fig. 4).

Fig. 4 Representation of the
melody in Fig. 2 without the
melodic figurations

4   Hidden Markov Model (HMM)

The Markov chains are a stochastic process, characterized by Markov properties.

It is a mathematical tool according to which the probability of a certain future event to occur depends uniquely on the current state [11]. Let $X_n = 1$ be the current state and $X_{n+k} = j$ the state after k steps, with i, j belonging to the set of states.

The conditional probability $p[X_{n+k} = j|X_n = i]$ is called a transition probability in k steps of the Markov chain [7].

The probability of transitioning from state i to state j in k steps of a homogeneous chain is indicated by Eq. (1):

$$p_{i,j}(k) = P[X_{n+k} = j|X_n = i] \tag{1}$$

By tagging the states as 1, 2, ...$n + 1$ we can summarize all the transition probabilities, $p_{i,j}(k)$, in a matrix $P(k)$, of the dimension n × n, where in the jth column and the ith row there is the transition probability from state i to state j in k steps:

$$P(k) = \begin{pmatrix} P_{1,1}(k) & \cdots & P_{1,j}(k) & \cdots & P_{1,n}(k) \\ P_{2,1}(k) & \cdots & P_{2,j}(k) & \cdots & P_{2,n}(k) \\ \cdots & \cdots & \cdots & \cdots & \cdots \\ P_{i,1}(k) & \cdots & P_{i,j}(k) & \cdots & P_{i,n}(k) \\ \cdots & \cdots & \cdots & \cdots & \cdots \\ P_{n,1}(k) & \cdots & P_{n,j}(k) & \cdots & P_{n,n}(k) \end{pmatrix}$$

The matrix $P(k)$, for $k = 1$, performs a fundamental role in Markov's chains theory: it (known as the transition matrix) represents the probability of transitioning to the next consecutive step [12].

It is possible to represent the transition matrix P by means of a graph called transitions diagram [13]. The latter consists in a graph the nodes of which represent the single states while the arcs, oriented and labeled with the probability, indicate the possible transitions [14]. For instance, considering the matrix

Fig. 5 State-transitions in a
Hidden Markov model

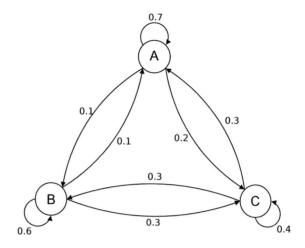

$$P(k) = \begin{pmatrix} 0.7 & 0.1 & 0.2 \\ 0.1 & 0.6 & 0.3 \\ 0.3 & 0.3 & 0.4 \end{pmatrix}$$

it has the following corresponding diagram (Fig. 5):

The problem of classification of the sequences may be solved by calculating the probability of the single sequence s to be emitted by the model M: $P(s|M)$. Formally (2):

$$P(s|M) = \sum \pi P(s, \pi|M) \qquad (2)$$

The designed algorithm uses a matrix of the transitions to construct a compositional logic able to create a musical idea [11]: the matrix represents the probabilities for a type of harmonic function to resolve to another type of harmonic function (Fig. 6).

A first and main task of the algorithm is to read music compositions in MIDI format (by different authors and of different ages), recognize the harmonic functions of the different musical degrees [15] and update the matrix of transitions.

Fig. 6 Matrix of the
transitions of the harmonic
functions

	T	S	D
T			
S			
D			

AWAIT

By reading an ever bigger number of music compositions, the algorithm will be able to propose ever more pleasant musical ideas: and this is because, by reading the mu- sic compositions, the probabilities of transition, but also the individual "*state- transitions*" (T, S, D) change, i.e. if a new harmonic function is identified (for in- stance the function Sp) this function will automatically be inserted in the matrix as a new "*state*" generating new transition probabilities.

5   The Results Obtained

The model of analysis set forth in this article was verified by realizing an algorithm the structure of which takes, most of all, in consideration each and every single aspect described above: the algorithm does not provide for any limit with respect to the dimensions of the transition matrix, but, on the contrary, it will be automatically dimensioned every time a new composition (already existing) is read, based on the characteristics of the respective composition.

The algorithm has the objective of proposing a new tonal musical idea as a source of inspiration for a new composition. As such, the new idea will have the typical characteristic of a musical phrase, i.e. it will not contain modulations (passage from one tonality to another) and the first harmonic function will be the tonic.

As far as the rhythmic structure is concerned, it has been decided to see to it that the new idea be representative of a harmonic structure (as the example in Fig. 4) that the composer might refine at a later stage, by inserting melodic figurations. Hence, a function, ergo a single sound, shall correspond to every single movement, bestowing a homorhythmic character on the melody (every movement will have the same duration).

The only parameters required as input for the elaboration are:

1. the musical tempo (a fractional number placed always at the beginning of the staff next to the key that indicates the total sum of the movements that must be contained in a beat and determines the sequence of the accents inside the same beat);
2. the number of beats (that the new idea will have to have).

On the basis of these parameters it is possible to define the total number of movements that will compose the new idea and, on the basis of the transition percent- ages derived from the transition matrix, we will determine for every function the number of times it will have to be repeated within the idea.

An example of a musical idea in a ¾ and four beats, generated after the reading of only three music compositions by different authors and different ages is illus- trated below (Fig. 7):

Fig. 7 Example of functional analysis and the related transition matrix updated after the reading of every music composition

- Theme of the melody "*Ah, vous dirai-je Maman*", KV 265;
- Moment musical No. 3 in F minor by Schubert;
- Song without words No. 9 by Mendelssohn.

For simplicity's sake in the demonstration and in order to better exploit the efficiency of the method, only the three main harmonic functions (T, S and D) were taken into consideration and all the secondary harmonic functions were ascribed to them (Sp counts as S, Dp counts as D....).

Furthermore and also solely for demonstrative purposes, only the first four beats of every composition were taken into consideration: it is important to specify that this choice was not motivated by the fact that the choice was made to create a four-beat phrase.

By means of the last transition matrix the algorithm determines the transition percentage from one state to the other (Fig. 8a) and therefore, on the basis of the total number of movements re w musical idea (in this example there are 12 because 4 beats of 3 are required), the number of times every function may occur (Fig. 8b).

(a)

	T	S	D
T	44	11	8
S	3	6	8
D	17		6

(b)

	T	S	D
T	5	1	1
S		1	1
D	2		1

Fig. 8 Representation of the transition percentages from one state to the other (**a**) and of the number of every function within the musical idea

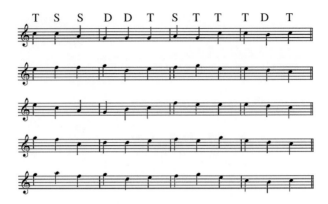

Fig. 9 Example of functional structure and related possible melodies

The results of Fig. 8b represent the basis for the random generation of the harmonic functions of the new idea. It is immediately deduced that there won't be a unique possible combination of harmonic functions, but, on the contrary, many different combinations may be obtained. The only common element of all these combinations is that the first harmonic function will always be the tonic one inasmuch as all the tonal music compositions always begin on the Tonic chord because it is representative of the main tonality.

In Fig. 9 below there is a representation of one of the possible combinations of harmonic functions and some possible examples of melodies, defined according to the rules of traditional harmony: every harmonic function is determined by the structure of the chord from which it derives and the chord is formed (fundamentally) by three sounds at a third distance one from the other.

In the example in Fig. 9 the main tonality is C major and therefore the harmonic functions will be represented by the following sounds:

1. T (representative of the first degree): C, E, G;
2. S (representative of the fourth degree): F, A, C;
3. D (representative of the fifth degree): G, B, D.

In this case, too, it is easy to understand how the presence in the transition matrix of the secondary harmonic functions may generate more appreciable melodies thanks to the presence of different combinations of sounds.

An example of how the third melody of Fig. 8 may be modified by the composer by using the melodic figurations is given in Fig. 10.

Fig. 10 Example of a melody

6   Conclusions

This article examined the use of Markov's process, as a mathematical means of information of the encoding with respect to the progression of the harmonic functions (as described by De la Motte) of the musical material. This mathematical process may be an efficient tool, used under the guidance of music theory, to formulate the models elaborated by the computer for the purposes of classic music composition.

The work presents several improvement opportunities. First of all, the possibility to consider, when reading harmonic structures in different music compositions, not only the functions of Tonic, Subdominant and Dominant, but also their correlated functions: the functions on degrees II, III, VI and VII.

Second of all, the possibility to incorporate in the proposed method the concept of "Cadence" which is very important on the compositional level for the definition of the musical phrase.

The tools presented in this article, developed on the basis of specifically musical objectives, are not meant in any way to be deemed a system for the composition of a musical piece, they rather represent a means of support to the didactic activity: a useful tool to allow specific in-depth analysis, stimulate the recovery of abilities that are not entirely acquired or as a simple tool of consultation and support to the explanation of the lecturer.

References

1. Cambouropoulos, E.: Markov chains as an aid to computer-assisted composition. Musical Prax. **1**(1), 41–52 (1994)
2. Wadi, A.: Analysis of music note patterns via markov chains, senior honors projects. Paper 2 (2012)
3. Lichtenwalter, R.N., Lichtenwalter, K., Chawla, N.V.: Applying learning algorithms to music generation. In: Proceedings of the 4th Indian International Conference on Artificial Intelligence, IICAI 2009
4. Cope, D.: Experiments in Musical Intelligence. A-R Editions Inc., Madison, WI (1996)
5. Chan, M., Potter, J., Schubert, E.: Improving algorithmic music composition with machine learning. In: Proceedings of the 9th International Conference on Music Perception and Cognition, ICMPC 2006
6. Lerdahl, F., Jackendoff, R.: A Generative Theory of Tonal Music. Cambridge: MIT Press (1983)
7. Moroni, A., Manzolli, J., Van Zuben, F., Gudwin, R., Populi, V.: An interactive evolutionary system for algorithmic music composition. Leonardo Music J **10**, 49–54 (2000)
8. Thom, B.: An interactive improvisational music companion, In: Proceeding of the Fourth International Conference on Autonomous Agents, pp 309–316. Barcelona, Spain (2000)
9. de la Motte, D.: Manuale di armonia, Bärenreiter (1976)
10. Coltro, B.: Lezioni di armonia complementare, Zanibon (1979)
11. Bengio, Y.: Markovian models for sequential data. Neural Comput Surv **2**, 129–162 (1999)
12. Bini, D.A., Latouche, G., Meini, B.: Numerical Methods for Structured Markov Chains. Oxford University Press (2005)

13. Rabiner, L.R.: A tutorial on hidden Markov models and selected applications in speech recognition. Proc. IEEE **77**(2), 257–286 (1989)
14. Kazi, N., Bhatia, S.: Various artificial intelligence techniques for automated melody generation. Int J Eng Res Technol **2**, 1646–1652 (2013)
15. Della Ventura, M.: Influence of the harmonic/functional analysis on the musical execution: representation and algorithm. In: Proceedings of the International Conference on Applied Informatics and Computing Theory, Barcellona (Spain) (2012)